Electrical Engineering

Electrical Engineering

Susan Curtis

Larsen & Keller
www.larsen-keller.com

Electrical Engineering
Susan Curtis
ISBN: 978-1-64172-656-6 (Hardback)

Larsen & Keller

Published by Larsen and Keller Education,
5 Penn Plaza,
19th Floor,
New York, NY 10001, USA

Cataloging-in-Publication Data

Electrical engineering / Susan Curtis.
p. cm.
Includes bibliographical references and index.
ISBN 978-1-64172-656-6
1. Electrical engineering. 2. Engineering. I. Curtis, Susan.
TK145 .E44 2022
621.3--dc23

For more information regarding Larsen and Keller Education and its products, please visit the publisher's website www.larsen-keller.com

Table of Contents

Preface

The purpose of this book is to help students understand the fundamental concepts of this discipline. It is designed to motivate students to learn and prosper. I am grateful for the support of my colleagues. I would also like to acknowledge the encouragement of my family.

The technical discipline which focuses on the equipment, systems and devices that use electromagnetism, electronics and electricity is referred to as electrical engineering. It is involved in their study, design and application. There are several sub-disciplines of electrical engineering. Some of them include telecommunications, computer engineering, signal processing, instrumentation engineering, electronic engineering, radio-frequency engineering and power engineering. Electrical engineering also deals with microelectronics and nanoelectronics. Microelectronics focuses on the designing of small electronic circuit components which are used in an integrated circuit. Nanoelectronics refers to the use of nanotechnology in electronic components. It deals with devices which are generally below 100 nanometers. This book presents the complex subject of electrical engineering in the most comprehensible and easy to understand language. Most of the topics introduced herein cover new techniques and applications of this field. This book is a complete source of knowledge on the present status of this important field.

A foreword for all the chapters is provided below:

Chapter - What is Electrical Engineering?

The branch of engineering that is concerned with the study, design and application of electronics, electricity and electromagnetism is referred to as electrical engineering. The major topics of study within this field include electrical power, electrical current and electrical resistance. This is an introductory chapter which will introduce briefly all the significant aspects of electrical engineering.

Chapter - Subfields of Electrical Engineering

Electrical engineering is a vast field that can be divided into various subfields. Some of them are power engineering, control engineering, telecommunications engineering and electronic engineering. The topics elaborated in this chapter will help in gaining a better perspective about these subfields of electrical engineering.

Chapter - Electrical Circuits

The closed path through which the electrical current flows is known as an electrical circuit. Some of the different types of electrical circuits are series circuits, parallel circuits, AC circuits and DC circuits. The chapter closely examines these types of electrical circuits to provide an extensive understanding of the subject.

Chapter - Electrical Laws and Circuit Theorems

Some of the major laws related to electrical circuits are Ohm's Law, Ampère's circuital law, Gauss's law, Coulomb's law, Faraday's laws of electromagnetic induction and Lenz's law. This chapter discusses in detail these laws related to electrical circuits as well as the different circuit theorems such as Superposition theorem, Thevenin's theorem, Norton's theorem, Maximum power transfer theorem etc.

Chapter - Tools and Instruments in Electrical Engineering

Various types of devices and equipment are used in electrical engineering. Some of them are LCR meter, voltmeter, ohmmeter, ammeter, current clamp, multimeter, network analyzer, power factor meter, etc. This chapter has been carefully written to provide an easy understanding of these tools and instruments used in electrical engineering.

Susan Curtis

1

What is Electrical Engineering?

The branch of engineering that is concerned with the study, design and application of electronics, electricity and electromagnetism is referred to as electrical engineering. The major topics of study within this field include electrical power, electrical current and electrical resistance. This is an introductory chapter which will introduce briefly all the significant aspects of electrical engineering.

Electrical engineering is one of the newer branches of engineering, and dates back to the late 19th century. It is the branch of engineering that deals with the technology of electricity. Electrical engineers work on a wide range of components, devices and systems, from tiny microchips to huge power station generators.

Early experiments with electricity included primitive batteries and static charges. However, the actual design, construction and manufacturing of useful devices and systems began with the implementation of Michael Faraday's Law of Induction, which essentially states that the voltage in a circuit is proportional to the rate of change in the magnetic field through the circuit. This law applies to the basic principles of the electric generator, the electric motor and the transformer. The advent of the modern age is marked by the introduction of electricity to homes, businesses and industry, all of which were made possible by electrical engineers.

Some of the most prominent pioneers in electrical engineering include Thomas Edison (electric light bulb), George Westinghouse (alternating current), Nikola Tesla (induction motor), Guglielmo Marconi(radio) and Philo T. Farnsworth (television). These innovators turned ideas and concepts about electricity into practical devices and systems that ushered in the modern age.

Since its early beginnings, the field of electrical engineering has grown and branched out into a number of specialized categories, including power generation and transmission systems, motors, batteries and control systems. Electrical engineering also includes electronics, which has itself branched into an even greater number of subcategories, such as radio frequency (RF) systems, telecommunications, remote sensing, signal processing, digital circuits, instrumentation, audio, video and optoelectronics.

The field of electronics was born with the invention of the thermionic valve diode vacuum tube in 1904 by John Ambrose Fleming. The vacuum tube basically acts as a current amplifier by outputting a multiple of its input current. It was the foundation of all electronics, including radios, television and radar, until the mid-20th century. It was largely supplanted by the transistor, which was developed in 1947 at AT&T's Bell Laboratories by William Shockley, John Bardeen and Walter Brattain, for which they received the 1956 Nobel Prize in physics.

Electricity

Electricity is a form of energy. Electricity is the flow of electrons. All matter is made up of atoms, and an atom has a center, called a nucleus. The nucleus contains positively charged particles called protons and uncharged particles called neutrons. The nucleus of an atom is surrounded by negatively charged particles called electrons. The negative charge of an electron is equal to the positive charge of a proton, and the number of electrons in an atom is usually equal to the number of protons. When the balancing force between protons and electrons is upset by an outside force, an atom may gain or lose an electron. When electrons are "lost" from an atom, the free movement of these electrons constitutes an electric current.

Electricity flows through the lightbulb filament, as a result the filament begins to glow and starts emitting light.

Electricity is a basic part of nature and it is one of our most widely used forms of energy. We get electricity, which is a secondary energy source, from the conversion of other sources of energy, like coal, natural gas, oil, nuclear power and other natural sources, which are called primary sources. Many cities and towns were built alongside waterfalls (a primary source of mechanical energy) that turned water wheels to perform work. Before electricity generation began slightly over 100 years ago, houses were lit with kerosene lamps, food was cooled in iceboxes, and rooms were warmed by wood-burning or coal-burning stoves. Beginning with Benjamin Franklin's experiment with a kite one stormy night in Philadelphia, the principles of electricity gradually became understood. In the mid-1800s, everyone's life changed with the invention of the electric light bulb. Prior to 1879, electricity had been used in arc lights for outdoor lighting. The lightbulb's invention used electricity to bring indoor lighting to our homes.

How is a Transformer Used?

To solve the problem of sending electricity over long distances, George Westinghouse developed a device called a transformer. The transformer allowed electricity to be efficiently transmitted over long distances. This made it possible to supply electricity to homes and businesses located far from the electric generating plant.

Despite its great importance in our daily lives, most of us rarely stop to think what life would be like without electricity. Yet like air and water, we tend to take electricity for granted. Everyday, we use electricity to do many functions for us -- from lighting and heating/cooling our homes, to being the power source for televisions and computers. Electricity is a controllable and convenient form of energy used in the applications of heat, light and power.

Today, the United States (U.S.) electric power industry is set up to ensure that an adequate supply of electricity is available to meet all demand requirements at any given instant.

How is Electricity Generated?

An electric generator is a device for converting mechanical energy into electrical energy. The process is based on the relationship between magnetism and electricity. When a wire or any other electrically conductive material moves across a magnetic field, an electric current occurs in the wire. The large generators used by the electric utility industry have a stationary conductor. A magnet attached to the end of a rotating shaft is positioned inside a stationary conducting ring that is wrapped with a long, continuous piece of wire. When the magnet rotates, it induces a small electric current in each section of wire as it passes. Each section of wire constitutes a small, separate electric conductor. All the small currents of individual sections add up to one current of considerable size. This current is what is used for electric power.

How are Turbines used to Generate Electricity?

An electric utility power station uses either a turbine, engine, water wheel, or other similar machine to drive an electric generator or a device that converts mechanical or chemical energy to electricity. Steam turbines, internal-combustion engines, gas combustion turbines, water turbines, and wind turbines are the most common methods to generate electricity.

Most of the electricity in the United States is produced in steam turbines. A turbine converts the kinetic energy of a moving fluid (liquid or gas) to mechanical energy. Steam turbines have a series of blades mounted on a shaft against which steam is forced, thus rotating the shaft connected to the generator. In a fossil-fueled steam turbine, the fuel is burned in a furnace to heat water in a boiler to produce steam.

Coal, petroleum (oil), and natural gas are burned in large furnaces to heat water to make steam that in turn pushes on the blades of a turbine. Did you know that coal is the largest single primary source of energy used to generate electricity in the United States? In 1998, more than half (52%) of the county's 3.62 trillion kilowatt-hours of electricity used coal as its source of energy.

Natural gas, in addition to being burned to heat water for steam, can also be burned to produce hot combustion gases that pass directly through a turbine, spinning the blades of the turbine to generate electricity. Gas turbines are commonly used when electricity utility usage is in high demand. In 1998, 15% of the nation's electricity was fueled by natural gas.

Petroleum can also be used to make steam to turn a turbine. Residual fuel oil, a product refined from crude oil, is often the petroleum product used in electric plants that use petroleum to make steam. Petroleum was used to generate less than three percent (3%) of all electricity generated in U.S. electricity plants in 1998.

Nuclear power is a method in which steam is produced by heating water through a process called nuclear fission. In a nuclear power plant, a reactor contains a core of nuclear fuel, primarily enriched uranium. When atoms of uranium fuel are hit by neutrons they fission (split), releasing heat and more neutrons. Under controlled conditions, these other neutrons can strike more uranium atoms, splitting more atoms, and so on. Thereby, continuous fission can take place, forming a chain reaction releasing heat. The heat is used to turn water into steam, that, in turn, spins a turbine that generates electricity. In 2015, Nuclear power is used to generate 19.47 percent of all the country's electricity.

As of 2013, hydropower accounts for 6.8 percent of U.S. electricity generation. Its a process in which flowing water is used to spin a turbine connected to a generator. There are mainly two basic types of hydroelectric systems that produce electricity. In the first system, flowing water accumulates in reservoirs created by the use of dams. The water falls through a pipe called a penstock and applies pressure against the turbine blades to drive the generator to produce electricity. In the second system, called run-of-river, the force of the river current (rather than falling water) applies pressure to the turbine blades to produce electricity.

Other Generating Sources

Geothermal power comes from heat energy buried beneath the surface of the earth. In some areas of the country, magma (molten matter under the earth's crust) flows close enough to the surface of the earth to heat underground water into steam, which can be tapped for use at steam-turbine plants. As of 2013, this energy source generates less than 1% of the electricity in the country, though an assessment by the U.S. Energy Information Administration that nine western states can potentially produce enough electricity to supply 20 percent of the nation's energy needs.

Solar power is derived from the energy of the sun. However, the sun's energy is not available full-time and it is widely scattered. The processes used to produce electricity using the sun's energy have historically been more expensive than using conventional fossil fuels. Photovoltaic conversion generates electric power directly from the light of the sun in a photovoltaic (solar) cell. Solar-thermal electric generators use the radiant energy from the sun to produce steam to drive turbines. In 2015, less than 1% of the nation's electricity was supplied by solar power.

Wind power is derived from the conversion of the energy contained in wind into electricity. Wind power, like the sun, is usually an expensive source of producing electricity. In 2014, It was used for roughly 4.44 percent of the nation's electricity. A wind turbine is similar to a typical wind mill.

Biomass (wood, municipal solid waste (garbage), and agricultural waste, such as corn cobs and wheat straw, are some other energy sources for producing electricity. These sources replace fossil fuels in the boiler. The combustion of wood and waste creates steam that is typically used in conventional steam-electric plants. In 2015, biomass accounts for 1.57 percent of the electricity generated in the United States.

The electricity produced by a generator travels along cables to a transformer, which changes electricity from low voltage to high voltage. Electricity can be moved long distances more efficiently using high voltage. Transmission lines are used to carry the electricity to a substation. Substations have transformers that change the high voltage electricity into lower voltage electricity. From the substation, distribution lines carry the electricity to homes, offices and factories, which require low voltage electricity.

How is Electricity Measured?

Electricity is measured in units of power called watts. It was named to honor James Watt, the inventor of the steam engine. One watt is a very small amount of power. It would require nearly 750 watts to equal one horsepower. A kilowatt represents 1,000 watts. A kilowatt-hour (kWh) is equal to the energy of 1,000 watts working for one hour. The amount of electricity a power plant generates or a customer uses over a period of time is measured in kilowatt-hours (kWh). Kilowatt-hours are determined by multiplying the number of kW's required by the number of hours of use. For example, if you use a 40-watt light bulb 5 hours a day, you have used 200 watts of power, or.2 kilowatt-hours of electrical energy.

Electromagnetism

Electromagnetism is the science of charge and of the forces and fields associated with charge. Electricity and magnetism are two aspects of electromagnetism.

Electricity and magnetism were long thought to be separate forces. It was not until the 19th century that they were finally treated as interrelated phenomena. In 1905 Albert Einstein's special theory of relativity established beyond a doubt that both are aspects of one common phenomenon. At a practical level, however, electric and magnetic forces behave quite differently and are described by different equations. Electric forces are produced by electric charges either at rest or in motion. Magnetic forces, on the other hand, are produced only by moving charges and act solely on charges in motion.

Electric phenomena occur even in neutral matter because the forces act on the individual charged constituents. The electric force in particular is responsible for most of the physical and chemical properties of atoms and molecules. It is enormously strong compared with gravity. For example, the absence of only one electron out of every billion molecules in two 70-kilogram (154-pound) persons standing two metres (two yards) apart would repel them with a 30,000-ton force. On a more familiar scale, electric phenomena are responsible for the lightning and thunder accompanying certain storms.

Electric and magnetic forces can be detected in regions called electric and magnetic fields. These fields are fundamental in nature and can exist in space far from the charge or current that generated them. Remarkably, electric fields can produce magnetic fields and vice versa, independent of any external charge. A changing magnetic field produces an electric field, as the English physicist Michael Faraday discovered in work that forms the basis of electric power generation. Conversely, a changing electric field produces a magnetic field, as the Scottish physicist James Clerk Maxwell deduced. The mathematical equations formulated by Maxwell incorporated light and wave phenomena into electromagnetism. He showed that electric and magnetic fields travel together through space as waves of electromagnetic radiation, with the changing fields mutually sustaining each other. Examples of electromagnetic waves traveling through space independent of matter are radio and television waves, microwaves, infrared rays, visible light, ultraviolet light, X-rays, and gamma rays. All of these waves travel at the same speed—namely, the velocity of light (roughly 300,000 kilometres, or 186,000 miles, per second). They differ from each other only in the frequency at which their electric and magnetic fields oscillate.

Maxwell's equations still provide a complete and elegant description of electromagnetism down to, but not including, the subatomic scale. The interpretation of his work, however, was broadened in the 20th century. Einstein's special relativity theory merged electric and magnetic fields into one common field and limited the velocity of all matter to the velocity of electromagnetic radiation. During the late 1960s, physicists discovered that other forces in nature have fields with a mathematical structure similar to that of the electromagnetic field. These other forces are the strong force, responsible for the energy released in nuclear fusion, and the weak force, observed in the radioactive decay of unstable atomic nuclei. In particular, the weak and electromagnetic forces have been combined into a common force called the electroweak force. The goal of many physicists to unite all of the fundamental forces, including gravity, into one grand unified theory has not been attained to date.

An important aspect of electromagnetism is the science of electricity, which is concerned with the behaviour of aggregates of charge, including the distribution of charge within matter and the motion of charge from place to place. Different types of materials are classified as either conductors or insulators on the basis of whether charges can move freely through their constituent matter. Electric current is the measure of the flow of charges; the laws governing currents in matter are important in technology, particularly in the production, distribution, and control of energy.

The concept of voltage, like those of charge and current, is fundamental to the science of electricity. Voltage is a measure of the propensity of charge to flow from one place to another; positive charges generally tend to move from a region of high voltage to a region of lower voltage. A common problem in electricity is determining the relationship between voltage and current or charge in a given physical situation.

This topic seeks to provide a qualitative understanding of electromagnetism as well as a quantitative appreciation for the magnitudes associated with electromagnetic phenomena.

Fundamentals

Everyday modern life is pervaded by electromagnetic phenomena. When a light bulb is switched on, a current flows through a thin filament in the bulb, and the current heats the filament to such a high temperature that it glows, illuminating its surroundings. Electric clocks and connections link simple devices of this kind into complex systems such as traffic lights that are timed and synchronized with the speed of vehicular flow. Radio and television sets receive information carried by electromagnetic waves traveling through space at the speed of light. To start an automobile, currents in an electric starter motor generate magnetic fields that rotate the motor shaft and drive engine pistons to compress an explosive mixture of gasoline and air; the spark initiating the combustion is an electric discharge, which makes up a momentary current flow.

Coulomb's Law

Many of these devices and phenomena are complex, but they derive from the same fundamental laws of electromagnetism. One of the most important of these is Coulomb's law, which describes the electric force between charged objects. Formulated by the 18th-century French physicist Charles-Augustin de Coulomb, it is analogous to Newton's law for the gravitational force. Both gravitational and electric forces decrease with the square of the distance between the objects, and both forces act along a line between them. In Coulomb's law, however, the magnitude and sign of the electric force are determined by the charge, rather than the mass, of an object. Thus, charge determines how electromagnetism influences the motion of charged objects. (Charge is a basic property of matter. Every constituent of matter has an electric charge with a value that can be positive, negative, or zero. For example, electrons are negatively charged, and atomic nuclei are positively charged. Most bulk matter has an equal amount of positive and negative charge and thus has zero net charge.)

According to Coulomb, the electric force for charges at rest has the following properties:

- Like charges repel each other, and unlike charges attract. Thus, two negative charges repel one another, while a positive charge attracts a negative charge.
- The attraction or repulsion acts along the line between the two charges.
- The size of the force varies inversely as the square of the distance between the two charges. Therefore, if the distance between the two charges is doubled, the attraction or repulsion becomes weaker, decreasing to one-fourth of the original value. If the charges come 10 times closer, the size of the force increases by a factor of 100.
- The size of the force is proportional to the value of each charge. The unit used to measure charge is the coulomb (C). If there were two positive charges, one of 0.1 coulomb and the second of 0.2 coulomb, they would repel each other with a force that depends on the product 0.2×0.1. If each of the charges were reduced by one-half, the repulsion would be reduced to one-quarter of its former value.

Static cling is a practical example of the Coulomb force. In static cling, garments made of synthetic material collect a charge, especially in dry winter air. A plastic or rubber comb passed quickly through hair also becomes charged and will pick up bits of paper. The synthetic fabric and the comb are insulators; charge on these objects cannot move easily from one part of the object to another. Similarly, an office copy machine uses electric force to attract particles of ink to paper.

Principle of Charge Conservation

Like Coulomb's law, the principle of charge conservation is a fundamental law of nature. According to this principle, the charge of an isolated system cannot change. If an additional positively charged particle appears within a system, a particle with a negative charge of the same magnitude will be created at the same time; thus, the principle of conservation of charge is maintained. In nature, a pair of oppositely charged particles is created when high-energy radiation interacts with matter; an electron and a positron are created in a process known as pair production.

The smallest subdivision of the amount of charge that a particle can have is the charge of one proton, $+1.602 \times 10^{-19}$ coulomb. The electron has a charge of the same magnitude but opposite sign—i.e., -1.602×10^{-19} coulomb. An ordinary flashlight battery delivers a current that provides a total charge flow of approximately 5,000 coulomb, which corresponds to more than 10^{22} electrons, before it is exhausted.

Electric current is a measure of the flow of charge, as, for example, charge flowing through a wire. The size of the current is measured in amperes and symbolized by i. An ampere of current represents the passage of one coulomb of charge per second, or 6.2 billion billion electrons (6.2×10^{18} electrons) per second. A current is positive when it is in the direction of the flow of positive charges; its direction is opposite to the flow of negative charges.

Electric Fields and Forces

The force and conservation laws are only two aspects of electromagnetism, however. Electric and magnetic forces are caused by electromagnetic fields. The term *field* denotes a property of space, so that the field quantity has a numerical value at each point of space. These values may also vary with time. The value of the electric or magnetic field is a vector—i.e., a quantity having both magnitude and direction. The value of the electric field at a point in space, for example, equals the force that would be exerted on a unit charge at that position in space.

Every charged object sets up an electric field in the surrounding space. A second charge "feels" the presence of this field. The second charge is either attracted toward the initial charge or repelled from it, depending on the signs of the charges. Of course, since the second charge also has an electric field, the first charge feels its presence and is either attracted or repelled by the second charge too.

The electric field from a charge is directed away from the charge when the charge is positive and toward the charge when it is negative. The electric field from a charge at rest is shown in figure for various locations in space. The arrows point in the direction of the electric field, and the length of the arrows indicates the strength of the field at the midpoint of the arrows.

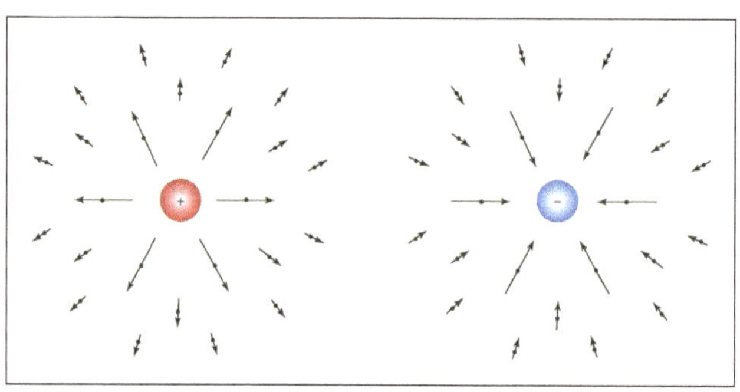

Electric fields. (Left) Field of a positive electric charge and (right) field of a negative electric charge.

If a positive charge were placed in the electric field, it would feel a force in the direction of the field. A negative charge would feel a force in the direction opposite the direction of the field.

In calculations, it is often more convenient to deal directly with the electric field than with the charges. Frequently, more is known about the field than about the distribution of charges in space.

For example, the distribution of charges in conductors is generally unknown because the charges move freely within the conductor. In static situations, however, the electric field in a conductor in equilibrium has a definite value, zero, because any force on the charges inside the conductor redistributes them until the field vanishes. The unit of electric field is newtons per coulomb, or volts per metre.

The electric potential is another useful field. It provides an alternative to the electric field in electrostatics problems. The potential is easier to use, however, because it is a single number, a scalar, instead of a vector. The difference in potential between two places measures the degree to which charges are influenced to move from one place to another. If the potential is the same at two places (i.e., if the places have the same voltage), charges will not be influenced to move from one place to the other. The potential on an object or at some point in space is measured in volts; it equals the electrostatic energy that a unit charge would have at that position. In a typical 12-volt car battery, the battery terminal that is marked with a + sign is at a potential 12 volts greater than the potential of the terminal marked with the – sign. When a wire, such as the filament of a car headlight, is connected between the + and the – terminals of the battery, charges move through the filament as an electric current and heat the filament, and the hot filament radiates light.

Magnetic Fields and Forces

The magnetic force influences only those charges that are already in motion. It is transmitted by the magnetic field. Both magnetic fields and magnetic forces are more complicated than electric fields and electric forces. The magnetic field does not point along the direction of the source of the field; instead, it points in a perpendicular direction. In addition, the magnetic force acts in a direction that is perpendicular to the direction of the field. In comparison, both the electric force and the electric field point directly toward or away from the charge.

The present discussion will deal with simple situations in which the magnetic field is produced by a current of charge in a wire. Certain materials, such as copper, silver, and aluminum, are conductors that allow charge to flow freely from place to place. If an external influence establishes a current in a conductor, the current generates a magnetic field. For a long straight wire, the magnetic field has a direction that encircles the wire on a plane perpendicular to the wire. The strength of the magnetic field decreases with distance from the wire. The arrows in figure represent the size and direction of the magnetic field for a current moving in the direction indicated. Figure A shows an end view with the current coming toward the reader, while figure B provides a three-dimensional view of the magnetic field at one position along the wire.

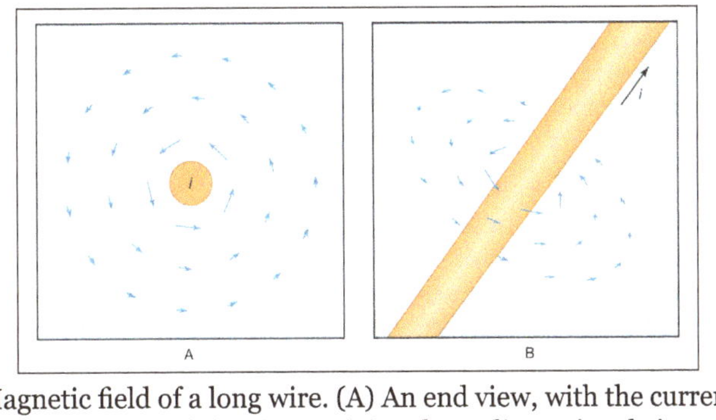

Magnetic field of a long wire. (A) An end view, with the current flowing toward the viewer. (B) A three-dimensional view.

In subsequent figures, continuous lines will be used to represent the direction of electric and magnetic fields. These lines emphasize the important fact that electric fields begin on positive charges and end on negative charges, while magnetic fields do not have beginnings or ends and close on themselves. The magnetic field shown in figure is unusually simple. Highly complex and useful magnetic fields can be generated by the proper choice of conductors to carry electric currents. Under development are thermonuclear fusion reactors for obtaining energy from the fusion of light nuclei in the form of very hot plasmas of hydrogen isotopes. The plasmas have to be confined by magnetic fields (dubbed "magnetic bottles") as no material container can withstand such high temperatures. Charged particles are also confined by magnetic fields in nature. Large numbers of charged particles, mostly protons and electrons, are trapped in huge bands around Earth by its magnetic field. These bands are known as the Van Allen radiation belts. Disturbance of Earth's confining magnetic field produces spectacular displays, the so-called northern lights, in which trapped charged particles are freed and crash through the atmosphere to Earth.

Interaction of a Magnetic Field with a Charge

How does the magnetic field interact with a charged object? If the charge is at rest, there is no interaction. If the charge moves, however, it is subjected to a force, the size of which increases in direct proportion with the velocity of the charge. The force has a direction that is perpendicular both to the direction of motion of the charge and to the direction of the magnetic field. There are two possible precisely opposite directions for such a force for a given direction of motion. This apparent ambiguity is resolved by the fact that one of the two directions applies to the force on a moving positive charge while the other direction applies to the force on a moving negative charge. Figure illustrates the directions of the magnetic force on positive charges and on negative charges as they move in a magnetic field that is perpendicular to the motion.

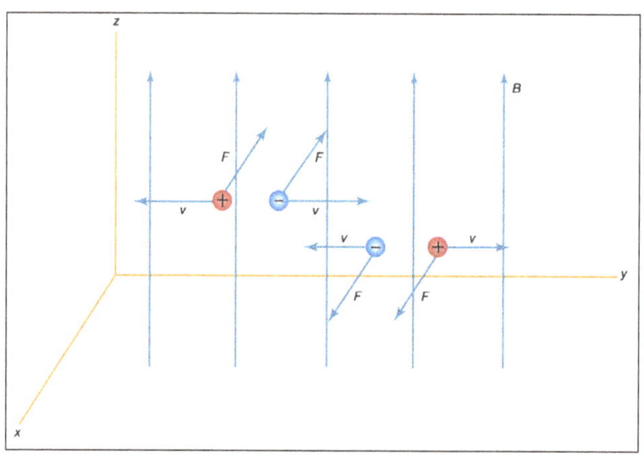

Magnetic force on moving charges. The magnetic force F is proportional to the charge and to the magnitude of velocity v times the magnetic field B.

Depending on the initial orientation of the particle velocity to the magnetic field, charges having a constant speed in a uniform magnetic field will follow a circular or helical path.

Electric currents in wires are not the only source of magnetic fields. Naturally occurring minerals exhibit magnetic properties and have magnetic fields. These magnetic fields result from the motion of electrons in the atoms of the material. They also result from a property of electrons called the magnetic dipole moment, which is related to the intrinsic spin of individual electrons. In most materials, little or no field is observed outside the matter because of the random orientation of the various constituentatoms. In some materials such as iron, however, atoms within certain distances tend to become aligned in one particular direction.

Magnets have numerous applications, ranging from use as toys and paper holders on home refrigerators to essential components in electric generators and machines that can accelerate particles to speeds approaching that of light. The practical application of magnetism in technology is greatly enhanced by using iron and other ferromagneticmaterials with electric currents in devices like motors. These materials amplify the magnetic field produced by the currents and thereby create more powerful fields.

While electric and magnetic effects are well separated in many phenomena and applications, they are coupled closely together when there are rapid time fluctuations. Faraday's law of induction describes how a time-varying magnetic field produces an electric field. Important practical applications include the electric generator and transformer. In a generator, the physical motion of a magnetic field produces electricity for power. In a transformer, electric power is converted from one voltage level to another by the magnetic field of one circuit inducing an electric current in another circuit.

The existence of electromagnetic waves depends on the interaction between electric and magnetic fields. Maxwell postulated that a time-varying electric field produces a magnetic field. His theory predicted the existence of electromagnetic waves in which each time-varying field produces the other field. For example, radio waves are generated by electronic circuits known as oscillators that cause rapidly oscillating currents to flow in antennas; the rapidly varying magnetic field has an associated varying electric field. The result is the emission of radio waves into space.

Many electromagnetic devices can be described by circuits consisting of conductors and other elements. These circuits may operate with a steady flow of current, as in a flashlight, or with time-varying currents. Important elements in circuits include sources of power called electromotive forces; resistors, which control the flow of current for a given voltage; capacitors, which store charge and energy temporarily; and inductors, which also store electrical energy for a limited time. Circuits with these elements can be described entirely with algebra.

Two mathematical quantities associated with vector fields, like the electric field E and the magnetic field B, are useful for describing electromagnetic phenomena. They are the flux of such a field through a surface and the line integral of the field along a path. The flux of a field through a surface measures how much of the field penetrates through the surface; for every small section of the surface, the flux is proportional to the area of that section and depends also on the relative orientation of the section and the field. The line integral of a field along a path measures the degree to which the field is aligned with the path; for every small section of path, it is proportional to the length of that section and is also dependent on the alignment of the field with that section of path. When the field is perpendicular to the path, there is no contribution to the line integral. The fluxes of E and B through a surface and the line integrals of these fields along a path play an important role in electromagnetic theory. As examples, the flux of the electric field E through a closed surface measures the amount of charge contained within the surface; the flux of the magnetic field B through a closed surface is always zero because there are no magnetic monopoles (magnetic charges consisting of a single pole) to act as sources of the magnetic field in the way that charge is a source of the electric field.

Effects of Varying Magnetic Fields

The merger of electricity and magnetism from distinct phenomena into electromagnetism is tied to three closely related events. The first was Hans Christian Ørsted's accidental discovery of the influence of an electric current on a magnetic needle—namely, that magnetic fields are produced by electric currents. Ørsted's 1820 report of his observation spurred an intense effort by scientists to prove that magnetic fields can induce currents. The second event was Faraday's experimental proof that a changing magnetic field can induce a current in a circuit. The third was Maxwell's prediction that a changing electric field has an associated magnetic field. The technological revolution attributed to the development of electric power and radio communications can be traced to these three landmarks.

Faraday's Law of Induction

Faraday's discovery in 1831 of the phenomenon of magnetic induction is one of the great milestones in the quest toward understanding and exploiting nature. Stated simply, Faraday found that (1) a changing magnetic field in a circuit induces an electromotive force in the circuit; and (2) the magnitude of the electromotive force equals the rate at which the flux of the magnetic field through the circuit changes. The flux is a measure of how much field penetrates through the circuit. The electromotive force is measured in volts and is represented by the equation,

$$emf = -\frac{d\Phi}{dt}.$$

Here, Φ, the flux of the vector field *B* through the circuit, measures how much of the field passes through the circuit. To illustrate the meaning of flux, imagine how much water from a steady rain will pass through a circular ring of area *A*. When the ring is placed parallel to the path of the water drops, no water passes through the ring. The maximum rate at which drops of rain pass through the ring occurs when the surface is perpendicular to the motion of the drops. The rate of water drops crossing the surface is the flux of the vector field ρv through that surface, where ρ is the density of water drops and *v* represents the velocity of the water. Clearly, the angle between *v* and the surface is essential in determining the flux. To specify the orientation of the surface, a vector *A* is defined so that its magnitude is the surface area *A* in units of square metres and its direction is perpendicular to the surface. The rate at which raindrops pass through the surface is $\rho v \cos \theta A$, where θ is the angle between *v* and *A*. Using vectornotation, the flux is $\rho v \cdot A$. For the magnetic field, the amount of flux through a small area represented by the vector *dA* is given by $B \cdot dA$. For a circuit consisting of a single turn of wire, adding the contributions from the entire surface that is surrounded by the wire gives the magnetic flux Φ of above equation. The rate of change of this flux is the induced electromotive force. The units of magnetic flux are webers, with one weber equaling one tesla per square metre. Finally, the minus sign in above equation indicates the direction of the induced electromotive force and hence of any induced current. The magnetic flux through the circuit generated by the induced current is in whatever direction will keep the total flux in the circuit from changing. The minus sign in above equation is an example of Lenz's law for magnetic systems. This law, deduced by the Russian-born physicist Heinrich Friedrich Emil Lenz, states that "what happens is that which opposes any change in the system."

$$emf = -\frac{d\Phi}{dt}.$$

Faraday's law is valid regardless of the process that causes the magnetic flux to change. It may be that a magnet is moved closer to a circuit or that a circuit is moved closer to a magnet. Figure shows a magnet brought near a conducting ring and gives the direction of the induced current and field, thus illustrating both Faraday's and Lenz's laws. Another alternative is that the circuit may change in size in a fixed external magnetic field or, as in the case of alternating-current (AC) generation, that the circuit may be a coil of conducting wire rotating in a magnetic field so that the flux Φ varies sinusoidally in time.

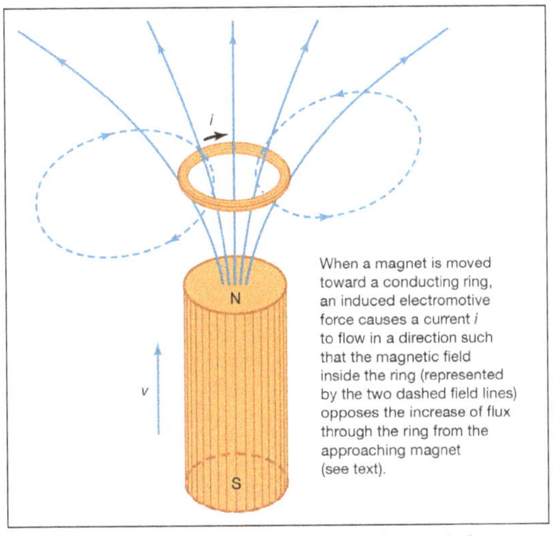

Demonstration of Faraday's and Lenz's laws.

The magnetic flux Φ through a circuit has to be considered carefully in the application of Faraday's law given in equation $emf = -\frac{d\Phi}{dt}$. For example, if a circuit consists of a coil with five closely spaced turns and if ϕ is the magnetic flux through a single turn, then the value of Φ for the five-turn circuit that must be used in Faraday's law is $\Phi = 5\phi$. If the five turns are not the same size and closely spaced, the problem of determining Φ can be quite complex.

Self-inductance and Mutual Inductance

The self-inductance of a circuit is used to describe the reaction of the circuit to a changing current in the circuit, while the mutual inductance with respect to a second circuit describes the reaction to a changing current in the second circuit. When a current i_1 flows in circuit 1, i_1 produces a magnetic field B_1; the magnetic flux through circuit 1 due to current i_1 is Φ_{11}. Since B_1 is proportional to i_1, Φ_{11} is as well. The constant of proportionality is the self-inductance L_1 of the circuit. It is defined by the equation,

$$\Phi_{11} = L_{[i]}.$$

The units of inductance are henrys. If a second circuit is present, some of the field B_1 will pass through circuit 2 and there will be a magnetic flux Φ_{21}in circuit 2 due to the current i_1. The mutual inductance M_{21} is given by,

$$\Phi_{21} = M2_{[i]}$$

The magnetic flux in circuit 1 due to a current in circuit 2 is given by $\Phi_{12} = M_{12}i_2$. An important property of the mutual inductance is that $M_{21} = M_{12}$. It is therefore sufficient to use the label M without subscripts for the mutual inductance of two circuits.

The value of the mutual inductance of two circuits can range from $+\sqrt{L_1L_2}$ to $-\sqrt{L_1L_2}$, depending on the flux linkage between the circuits. If the two circuits are very far apart or if the field of one circuit provides no magnetic flux through the other circuit, the mutual inductance is zero. The maximum possible value of the mutual inductance of two circuits is approached as the two circuits produce B fields with increasingly similar spatial configurations.

If the rate of change with respect to time is taken for the terms on both sides of equation $\Phi_{11} = L_{[i]}$. the result is $d\Phi_{11}/dt = L_1 di_1/dt$. According to Faraday's law, $d\Phi_{11}/dt$ is the negative of the induced electromotive force. The result is the equation frequently used for a single inductor in an AC circuit—i.e.,

$$\Phi_{11} = L_{[i]}.$$

$$\text{emf} = -L\frac{di}{dt}.$$

The phenomenon of self-induction was first recognized by the American scientist Joseph Henry. He was able to generate large and spectacular electric arcs by interrupting the current in a large copper coil with many turns. While a steady current is flowing in a coil, the energy in the magnetic field is given by $^1/_2Li^2$. If both the inductance L and the current i are large, the amount of energy is also large. If the current is interrupted, as, for example, by opening a knife-blade switch, the current and therefore the magnetic flux through the coil drop quickly. Equation $\text{emf} = -L\frac{di}{dt}$. describes the resulting electromotive force induced in the coil, and a large potential difference is developed between the two poles of the switch. The energy stored in the magnetic field of the coil is dissipated as heat and radiation in an electric arc across the space between the terminals of the switch. Due to advances in superconducting wires for electromagnets, it is possible to use large magnets with magnetic fields of several teslas for temporarily storing electric energy as energy in the magnetic field. This is done to accommodate short-term fluctuations in the consumption of electric power.

$$\text{emf} = -L\frac{di}{dt}.$$

A transformer is an example of a device that uses circuits with maximum mutual induction. Figure illustrates the configuration of a typical transformer. Here, coils of insulated conducting wire are wound around a ring of iron constructed of thin isolated laminations or sheets. The laminations minimize eddy currents in the iron. Eddy currents are circulatory currents induced in the metal by the changing magnetic field. These currents produce an undesirable by-product—heat in the iron. Energy loss in a transformer can be reduced by using thinner laminations, very "soft" (low-carbon) iron and wire with a larger cross section, or by winding the primary and secondary circuits with conductors that have very low resistance. Unfortunately, reducing the heat loss increases the cost of transformers. Transformers used to transmit and distribute power are commonly 98 to 99 percent efficient. While eddy currents are a problem in transformers, they are useful for heating objects in a vacuum. Eddy currents are induced in the object to be heated by surrounding a relatively nonconducting vacuum enclosure with a coil carrying a high-frequency alternating current.

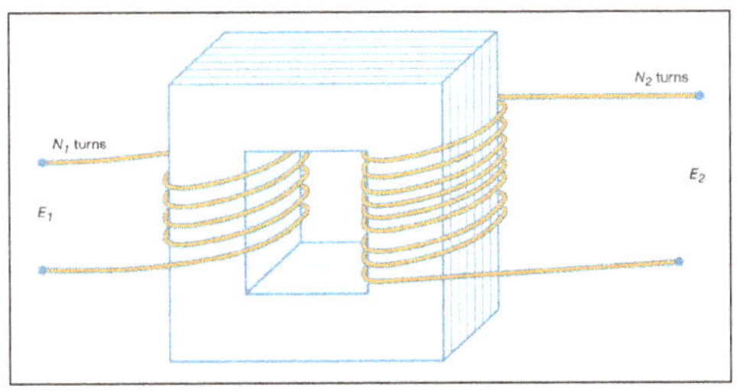

An AC transformer.

In a transformer, the iron ensures that nearly all the lines of B passing through one circuit also pass through the second circuit and that, in fact, essentially all the magnetic flux is confined to the iron. Each turn of the conducting coils has the same magnetic flux; thus, the total flux for each coil is proportional to the number of turns in the coil. As a result, if a source of sinusoidally varying electromotive force is connected to one coil, the electromotive force in the second coil is given by

$$\text{emf}_2 = \text{emf}_1 \frac{N_2}{N_1}.$$

Thus, depending on the ratio of N_2 to N_1 (where N_1 and N_2 are the number of turns in the first and second coils, respectively), the transformer can be either a step-up or a step-down device for alternating voltages. For many reasons, including safety, generation and consumption of electric power occur at relatively low voltages. Step-up transformers are used to obtain high voltages before electric power is transmitted, since for a given amount of power, the current in the transmission lines is much smaller. This minimizes energy lost by resistive heating of the conductors.

Faraday's law constitutes the basis for the power industry and for the transformation of mechanical energy into electric energy. In 1821, a decade before his discovery of magnetic induction, Faraday conducted experiments with electric wires rotating around compass needles. This earlier work, in which a wire carrying a current rotated around a magnetized needle and a magnetic needle was made to rotate around a wire carrying an electric current, provided the groundwork for the development of the electric motor.

Effects of Varying Electric Fields

Maxwell's prediction that a changing electric field generates a magnetic field was a masterstroke of pure theory. The Maxwell equations for the electromagnetic field unified all that was hitherto known about electricityand magnetism and predicted the existence of an electromagnetic phenomenon that can travel as waves with the velocity of 1/Square root of$\sqrt{\varepsilon_0 \mu_0}$ in a vacuum. That velocity, which is based on constants obtained from purely electric measurements, corresponds to the speed of light. Consequently, Maxwell concluded that light itself was an electromagnetic phenomenon. Later, Einstein's special relativity theory postulated that the value of the speed of light is independent of the motion of the source of the light. Since then, the speed of light has been measured with increasing accuracy. In 1983 it was defined to be exactly 299,792,458 metres per second. Together with the cesium clock, which has been used to define the second, the speed of light serves as the new standard for length.

The circuit in figure is an example of a magnetic field generated by a changing electric field. A capacitor with parallel plates is charged at a constant rate by a steady current flowing through the long, straight leads in figure A.

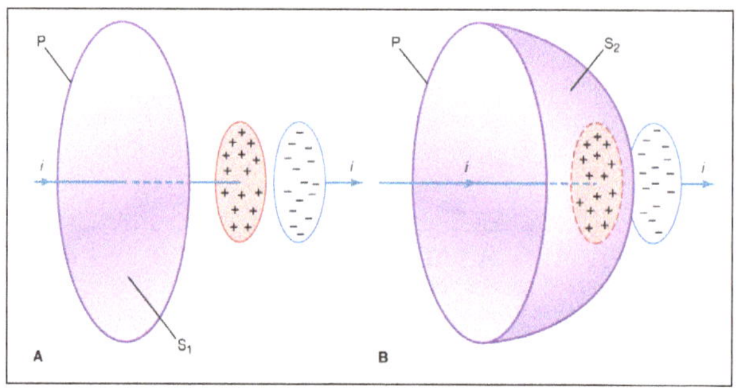

Current i charging a capacitor as an illustration of Maxwell's displacement current.

The objective is to apply Ampère's circuital law for magnetic fields to the path P, which goes around the wire in figure A. This law (named in honour of the French physicist André-Marie Ampère) can be derived from the Biot and Savart equation for the magnetic field produced by a current. Using vector calculus notation, Ampère's law states that the integral $\oint B{\cdot}dl$ along a closed path surrounding the current i is equal to $\mu_0 i$. (An integral is essentially a sum, and, in this case, $\oint B{\cdot}dl$ is the sum of $B \cos \theta dl$ taken for a small length of the path until the complete loop is included. At each segment of the path dl, θ is the angle between the field B and dl.) The current i in Ampère's law is the total flux of the current density J through any surface surrounded by the closed path. In figure A, the closed path is labeled P, and a surface S_1 is surrounded by path P. All the current density through S_1 lies within the conducting wire. The total flux of the current density is the current i flowing through the wire. The result for surface S_1 reflects the value of the magnetic field around the wire in the region of the path P. In figure B, path P is the same but the surface S_2 passes between the two plates of the capacitor. The value of the total flux of the current density through the surface should also be i. There is, however, clearly no motion of charge at all through the surface S_2. The dilemma is that the value of the integral $\oint B{\cdot}dl$ for the path P cannot be both $\mu_0 i$ and zero.

Maxwell's resolution of this dilemma was his conclusion that there must be some other kind of current density, called the displacement current J_d, for which the total flux through the surface S_2 would be the same as the current i through the surface S_1. J_d would take, for the surface S_2, the place of the current density J associated with the movement of charge, since J is clearly zero due to the lack of charges between the plates of the capacitor. What happens between the plates while the current i is flowing? Because the amount of charge on the capacitor increases with time, the electric field between the plates increases with time too. If the current stops, there is an electric field between the plates as long as the plates are charged, but there is no magnetic field around the wire. Maxwell decided that the new type of current density was associated with the changing of the electric field. He found that,

$$Jd = \frac{dD}{dt}.$$

where $D = \varepsilon_0 E$ and E is the electric field between the plates. In situations where matter is present, the field D is modified to include polarization effects; the result is $D = \varepsilon_0 E + P$. The field D is measured in coulombs per square metre. Adding the displacement current to Ampère's law represented Maxwell's prediction that a changing electric field also could be a source of the magnetic field B. Following Maxwell's predictions of electromagnetic waves, the German physicist Heinrich Hertz initiated the era of radio communications in 1887 by generating and detecting electromagnetic waves.

$$Jd = \frac{dD}{dt}.$$

Using vector calculus notation, the four equations of Maxwell's theory of electromagnetism are,

- $div\, D = \rho,$
- $div\, B = 0,$
- $curl\, E = -\frac{dB}{dt},$
- $curl\, H = J + \frac{dD}{dt},$

where $D = \varepsilon_0 E + P$, and $H = B/\mu_0 - M$. The first equation is based on Coulomb's inverse square law for the force between two charges; it is a form of Gauss's law, which relates the flux of the electric field through a closed surface to the total charge enclosed by the surface. The second equation is based on the fact that apparently no magnetic monopoles exist in nature; if they did, they would be point sources of magnetic field. The third is a statement of Faraday's law of magnetic induction, which reveals that a changing magnetic field generates an electric field. The fourth is Ampère's law as extended by Maxwell to include the displacement current discussed above; it associates a magnetic field to a changing electric field as well as to an electric current.

Maxwell's four equations represent a complete description of the classical theory of electromagnetism. His discovery that light is an electromagnetic wave meant that optics could be understood as part of electromagnetism. Only in microscopic situations is it necessary to modify Maxwell's equations to include quantum effects. That modification, known as quantum electrodynamics (QED), accounts for certain atomic properties to a degree of precision exceeding one part in 100 million.

Sometimes it is necessary to shield apparatus from external electromagnetic fields. For a static electric field, this is a simple matter; the apparatus is surrounded by a shield made of a good conductor (e.g., copper). Shielding apparatus from a steady magnetic field is more difficult because materials with infinite magnetic permeability μ do not exist; for example, a hollow shield made of soft iron will reduce the magnetic field inside to a considerable extent but not completely. It is sometimes possible to superpose a field in the opposite direction to produce a very low field region and then to use additional material with a high μ for shielding. In the case of electromagnetic waves, the penetration of the waves in matter varies, depending on the frequency of the radiation and the electric conductivity of the medium. The skin depth δ (which is the distance in the conducting medium traversed for an amplitude decrease of $1/e$, about 1/3) is given by,

$$\delta = \sqrt{\frac{2}{\omega \mu_0 \sigma_J}}.$$

At high frequency, the skin depth is small. Therefore, to transmit electronic messages through seawater, for example, a very low frequency must be used to get a reasonable fraction of the signal far below the surface.

A metal shield can have some holes in it and still be effective. For instance, a typical microwave oven has a frequency of 2.5 gigahertz, which corresponds to a wavelength of about 12 centimetres for the electromagnetic wave inside the oven. The metal shield on the door has small holes about two millimetres in diameter; the shield works because the wavelength of the microwave radiation is much greater than the size of the holes. On the other hand, the same shield is not effective with radiation of a much shorter wavelength. Visible light passes through the holes in the shield, as evidenced by the fact that it is possible to see inside a microwave oven when the door is closed.

Electric Current

Electric current is any movement of electric charge carriers, such as subatomic charged particles (e.g., electrons having negative charge, protons having positive charge), ions (atomsthat have lost or gained one or more electrons), or holes (electron deficiencies that may be thought of as positive particles).

Electric current in a wire, where the charge carriers are electrons, is a measure of the quantity of charge passing any point of the wire per unit of time. In alternating current the motion of the electric charges is periodically reversed; in direct current it is not. In many contexts the direction of the current in electric circuits is taken as the direction of positive charge flow, the direction opposite to the actual electron drift. When so defined the current is called conventional current.

Current is usually denoted by the symbol I. Ohm's law relates the current flowing through a conductor to the voltage V and resistance R; that is, $V = IR$. An alternative statement of Ohm's law is $I = V/R$.

Current in gases and liquids generally consists of a flow of positive ions in one direction together with a flow of negative ions in the opposite direction. To treat the overall effect of the current, its direction is usually taken to be that of the positive charge carrier. A current of negative charge moving in the opposite direction is equivalent to a positive charge of the same magnitude moving in the conventional direction and must be included as a contribution to the total current. Current in semiconductors consists of the motion of holes in the conventional direction and electrons in the opposite direction. Currents of many other kinds exist, such as beams of protons, positrons, or charged pions and muons in particle accelerators.

Electric current generates an accompanying magnetic field, as in electromagnets. When an electric current flows in an external magnetic field, it experiences a magnetic force, as in electric motors. The heat loss, or energy dissipated, by electric current in a conductor is proportional to the square of the current.

A common unit of electric current is the ampere, a flow of one coulomb of charge per second, or 6.2×10^{18} electrons per second. The centimetre–gram–second units of current is the electrostatic unit of charge (esu) per second. One amp equals 3×10^{9} esu per second.

Commercial power lines make available about 100 amps to a typical home; a 60-watt lightbulb pulls about 0.5 amp of current and a one-room air conditioner about 15 amps.

Electric Power

The rate at which the work is being done in an electrical circuit is called an electric power. In other words, the electric power is defined as the rate of the transferred of energy. The electric power is produced by the generator and can also be supplied by the electrical batteries. It gives a low entropy form of energy which is carried over long distance and also it is converted into various other forms of energy like motion, heat energy, etc.

The electric power is divided into two types, i.e., the AC power and the DC power. The classification of the electric power depends on the nature of the current. The electric power is sold regarding joule which is the product of the power in kilowatts and the running time of the machinery in hours. The utility of power is measured by the electric meter which records the total energy consumed by the powered devices. The electric power is given by the equation shown below.

$$\text{Electrical Power} = \frac{\text{work done in an electrical current}}{\text{time}}$$

$$P = \frac{VIt}{t} = VI = IR^2 = \frac{V^2}{R}$$

Where V is the voltage in volts, I is the current in amperes, R is the resistance offered by the powered devices, T is the time in seconds and the P is the power measured in watts.

Unit of Electric Power

The unit of electrical power is Watt.

If,

$$V = 1\,volts\ and\ I = 1\,ampere$$
$$P = 1\,watt$$

Thus, the power consumed in an electrical circuit is said to one watt if one ampere current flows through the circuit when a potential difference of 1 volt is applied across it.The bigger unit of electrical power is the kilowatt (kW), it is usually used in the power system,

$$1kW = 1000W$$

Types of an Electric Power

The electrical power is mainly classified into two types. They are the DC power and the AC power.

DC Power

The DC power is defined as the product of the voltage and current. It is produced by the fuel cell, battery and generator.

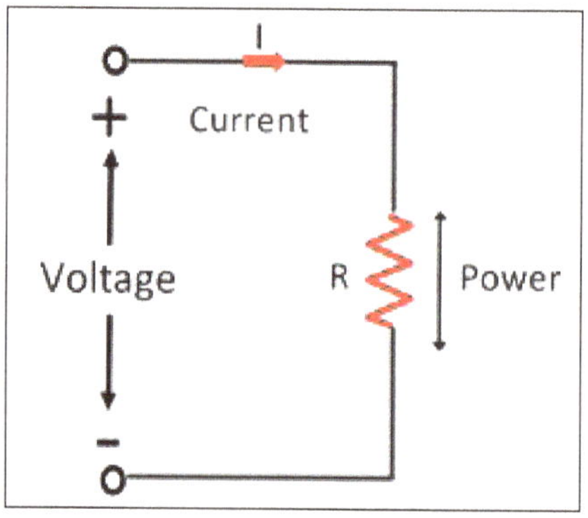

Electric Power.

$$P = V \times I$$

Where,

P – Power in watt.

V – voltage in volts.

I – current in amps.

AC Power

The AC power is mainly classified into three types. They are the apparent power, active power and real power.

1. Apparent Power – The apparent power is the useless power or idle power. It is represented by the symbol S, and their SI unit is volt-amp.

$$S = V_{rms} I_{rms}$$

Where,

S – apparent power

V_{rms} – RMS voltage = $V_{peak}\sqrt{2}$ in volt.

I_{rms} – RMS current = $I_{peak}\sqrt{2}$ in the amp.

2. Active Power – The active power (P) is the real power which is dissipated in the circuit resistance.

$$P = V_{max}\, I_{max}\, Cos\phi$$

Where,

P – the real power in watts.

V_{rms} – RMS voltage = $V_{peak}\sqrt{2}$ in volts.

I_{rms} – RMS current = $I_{peak}\sqrt{2}$ in the amp.

Φ – impedance phase angle between voltage and current.

3. Reactive Power – The power developed in the circuit reactance is called reactive power (Q). It is measured in volt-ampere reactive.

$$Q = V_{rms} I_{rms} Sin\phi$$

Where, Q – the reactive power in watts.

Vrms – RMS voltage = Vpeak√2 in volt.

Irms – RMS current = Ipeak√2 in the amp.

Φ – impedance phase angle between voltage and current.

The relation between the apparent, active and reactive power is shown below.

$$S^2 = Q^2 + P^2$$

The ratio of the real to the apparent power is called power factor, and their value lies between 0 and 1.

Electrical Units

Volt

Volt, a unit of electrical potential, is the potential difference and electromotive force in the metre–kilogram–second system (SI); it is equal to the difference in potential between two points in a conductor carrying one ampere current when the power dissipated between the points is one watt. An equivalent is the potential difference across a resistance of one ohm when one ampere is flowing through it. The volt is named in honour of the 18th–19th-century Italian physicist Alessandro

Volta. These units are defined in accordance with Ohm's law, that resistance equals the ratio of potential to current, and the respective units of ohm, volt, and ampere are used universally for expressing electrical quantities.

One Volt is defined as energy consumption of one joule per electric charge of one coulomb.

$$1V = 1J/C$$

One volt is equal to current of 1 amp times resistance of 1 ohm:

$$1V = 1A \cdot 1\Omega$$

Alessandro Volta

The Volt unit is named after Alessandro Volta, an Italian physicist who invented an electric battery.

name	symbol	conversion	example
microvolt	μV	$1\mu V = 10^{-6}V$	$V = 30\mu V$
millivolt	mV	$1mV = 10^{-3}V$	$V = 5mV$
volt	V	-	$V = 10V$
kilovolt	kV	$1kV = 10^{3}V$	$V = 2kV$
megavolt	MV	$1MV = 10^{6}V$	$V = 5MV$

Volts to Watts Conversion

The power in watts (W) is equal to the voltage in volts (V) times the current in amps (A):

watts (W) = volts (V) × amps (A)

Volts to Joules Conversion

The energy in joules (J) is equal to the voltage in volts (V) times the electric charge in coulombs (C):

joules (J) = volts (V) × coulombs (C)

Volts to Amps Conversion

The current in amps (A) is equal to the voltage in volts (V) divided by the resistance in ohms (Ω):

amps (A) = volts (V) / ohms(Ω)

The current in amps (A) is equal to the power in watts (W) divided by the voltage in volts (V):

amps (A) = watts (W) / volts (V)

Volts to Electron-volts Conversion

The energy in electronvolts (eV) is equal to the potential difference or voltage in volts (V) times the electric charge in electron charges (e):

electronvolts (eV) = volts (V) × electron-charge (e)

= volts (V) × 1.602176e-19 coulombs (C)

Ohm

Ohm, (Ω) is the unit of electrical resistance in the metre-kilogram-second system, named in honour of the 19th-century German physicist Georg Simon Ohm. It is equal to the resistance of a circuit in which a potential difference of one volt produces a current of one ampere ($1\Omega = 1\ V/A$); or, the resistance in which one watt of power is dissipated when one ampere flows through it. Ohm's law (*q.v.*) states that resistance equals the ratio of the potential difference to current, and the ohm, volt, and ampere are the respective fundamental units used universally for expressing quantities.

Impedance, the apparent resistance to an alternating current, and reactance, the part of impedance resulting from capacitance or inductance, are circuit characteristics that are measured in ohms. The acoustic ohm and the mechanical ohm are analogous units sometimes used in the study of acoustic and mechanical systems, respectively.

Watt

Watt is the unit of power (symbol: W).

The watt unit is named after James Watt, the inventor of the steam engine.

One watt is defined as the energy consumption rate of one joule per second.

1W = 1J / 1s

One watt is also defined as the current flow of one ampere with voltage of one volt.

1W = 1V × 1A

name	symbol	conversion	example
picowatt	pW	$1\text{pW} = 10^{-12}\text{W}$	P = 10 pW
nanowatt	nW	$1\text{nW} = 10^{-9}\text{W}$	P = 10 nW
microwatt	μW	$1\mu\text{W} = 10^{-6}\text{W}$	P = 10 μW
milliwatt	mW	$1\text{mW} = 10^{-3}\text{W}$	P = 10 mW
watt	W	-	P = 10 W
kilowatt	kW	$1\text{kW} = 10^{3}\text{W}$	P = 2 kW
megawatt	MW	$1\text{MW} = 10^{6}\text{W}$	P = 5 MW
gigawatt	GW	$1\text{GW} = 10^{9}\text{W}$	P = 5 GW

How to Convert Watts to Kilowatts

The power P in kilowatts (kW) is equal to the power P in watts (W) divided by 1000:

$$P_{(kW)} = P_{(W)} / 1000$$

How to Convert Watts to Milliwatts

The power P in milliwatts (mW) is equal to the power P in watts (W) times 1000:

$$P_{(mW)} = P_{(W)} \cdot 1000$$

How to Convert Watts to dBm

The power P in decibel-milliwatts (dBm) is equal to 10 times base 10 logarithm of the power P in milliwatts (mW) divided by 1 milliwatt:

$$P_{(dBm)} = 10 \cdot \log_{10}(P_{(mW)} / 1mW)$$

How to Convert Watts to Amps

The current I in ampers (A) is equal to the power P in watts (W) divided by the voltage V in volts (V):

$$I_{(A)} = P_{(W)} / V_{(V)}$$

How to Convert Watts to Volts

The voltage V in volts (V) is equal to the power P in watts (W) divided by the current I in ampers (A):

$$V_{(V)} = P_{(W)} / I_{(A)}$$

How to Convert Watts to Ohms

$$R_{(\Omega)} = P_{(W)} / I_{(A)}^2$$

$$R_{(\Omega)} = V_{(V)}^2 / P_{(W)}$$

How to Convert Watts to btu/hr

$$P_{(BTU/hr)} = 3.412142 \cdot P_{(W)}$$

How to Convert Watts to Joules

$$E_{(J)} = P_{(W)} \cdot t_{(s)}$$

How to Convert Watts to Horsepower

$$P_{(HP)} = P_{(W)} / 746$$

How to Convert Watts to kVA

The real power P in watts (W) is equal to 1000 times the apparent power S in kilo-volt-amperes (kVA) times the power factor (PF) or cosine of the phase angle φ:

$$P_{(W)} = 1000 \cdot S_{(kVA)} \cdot PF = 1000 \cdot S_{(kVA)} \cdot \cos \varphi$$

How to Convert Watts to VA

The real power P in watts (W) is equal to the apparent power S in volt-amperes (VA) times the power factor (PF) or cosine of the phase angle φ:

$$P_{(W)} = S_{(VA)} \cdot PF = S_{(VA)} \cdot \cos \varphi$$

Electrical Resistance

The electrical resistance is defined as the difficulty occurs in the flow of electrons. The conductor has free electrons. When the voltage or potential difference is applied across the conductor, the free electrons start moving in the particular direction. During the movement, these electrons collide with the atoms and molecules of the conductor. The atoms or molecules create the obstruction in the flow of electrons. This obstruction is called resistance.

The electrical resistance is provided to the circuit through the resistor. The resistance shows the relation between the applied voltage and the flow of charges. The resistance is inversely proportional to the flow of current.

Unit: Resistance is measured in ohms (kilo-ohms) and is denoted by symbols Ω (or kΩ). A wire is said to have a resistance of one ohm If the one-ampere current is passed through it.

Law of Resistance

Every material must possess resistance. The material which has good conductivity have low resistance whereas the material which has low conductivity has high resistance. The ohmmeter measures the electrical resistance. The ohmmeter is connected to the terminal of the circuit whose resistance is to be measured. The resistance of the wire depends on the following factors. These factors are:

- The resistance of a wire is directly proportional to its length, l, i.e., $R \propto l$.
- The resistance of the wire is inversely proportional to its area of cross section (a).

$$R = \rho \frac{l}{a}$$

$$\rho = \frac{Ra}{l}$$

Where ρ is a constant of proportionality also called the resistivity of the wire material. The values of ρ depend on the nature (atomic structure) of the materials.

- The resistance of the wire depends on the nature (atomic structure) of the material of which the wire is made.
- The resistance of the wire depends on the temperature of the wire.

Resistivity of the Material

The resistivity of the material is defined as the resistance offered by the one-meter length of wire (of given area) having an area of cross section of one square meter. The resistance of the wire is given as,

$$R \propto \frac{l}{a}$$

If, l = 1 meter; a = 1 m^2 then the resistance of the wire is R = ρ

In place of wire, if a cube of the one-meter side of a given material is taken and their two opposite faces are considered, then the resistance is given by the relation shown below.

$$i = 1m; a = 1 \times 1 = 1m^2 \; and \; R = \rho$$

Hence, the resistance offered between the two opposite two faces of the one-meter cube of the given material is called the resistivity of the material.

Unit: Resistance of the wire is given by the equation,

$$R \propto \frac{l}{a}$$

Also from the above equation, we get,

$$R = \rho \frac{l}{a}$$

$$\rho = \frac{Ra}{l}$$

Substituting the units of various quantities as per SI units, we get,

$$\rho = \frac{ohms \times m^2}{m} = ohm - m$$

Hence the unit of resistivity is ohm-meter in SI units.

Specific Resistance of the Material

The specific resistance or resistivity is the resistance of the unit length and the unit cross-section area of the material. It is measured in ohmmeter.

References

- Electrical-engineering: livescience.com, Retrieved 12 July, 2019
- What-is-electricity-4019643: thoughtco.com, Retrieved 11 January, 2019
- Electromagnetism, science: britannica.com, Retrieved 1 August, 2019
- Electric-current, science: britannica.com, Retrieved 5 March, 2019
- Electric-power: circuitglobe.com, Retrieved 8 June, 2019
- Volt-unit-of-measurement, science: britannica.com, Retrieved 3 May, 2019
- Volt, electric: rapidtables.com, Retrieved 5 February, 2019
- Ohm, science: britannica.com, Retrieved 30 April, 2019
- Watt, electric: rapidtables.com, Retrieved 15 July, 2019
- Electrical-resistance: circuitglobe.com, Retrieved 30 August, 2019

2
Subfields of Electrical Engineering

Electrical engineering is a vast field that can be divided into various subfields. Some of them are power engineering, control engineering, telecommunications engineering and electronic engineering. The topics elaborated in this chapter will help in gaining a better perspective about these subfields of electrical engineering.

Power Engineering

The very basic concepts of Power System Engineering were actually introduced in 1831 by Michael Faraday when he discovered the principle of Electromagnetic Induction. The world's first power station was built in 1881 at Godalming in England. It consisted of seven Siemens arc lamps that were supplied with AC with 250 volts. It was further advanced to Steam-Powered electric power station in 1882. It was called the "Pearl Street Station". The power industry kept on flourishing rapidly and many Power Companies came into being in a very short period of time in Europe and United States. Further developments were made in the 20 and the 21st century. The power systems were getting more efficient with passing time and have become a core area of electrical engineering field. What is Power Engineering – The Core of Electrical Engineering.

Power Transmission and Distribution / By Umair Mirza / Electrical Engineering.

Basics of Power Engineering

Power Engineering is actually a sub-field of engineering that relates to the following areas:

- Distribution, transmission & generation of electric power.
- The devices related to this field including transformers, electric generators and motors.
- AC and DC power and the conversion between them.
- The development of other power systems including aircraft.

The very basic concepts of Power System Engineering were actually introduced in 1831 by Michael Faraday when he discovered the principle of Electromagnetic Induction. The world's first power station was built in 1881 at Godalming in England. It consisted of seven Siemens arc lamps that were supplied with AC with 250 volts. It was further advanced to Steam-Powered electric power station in 1882. It was called the "Pearl Street Station". The power industry kept on flourishing rapidly and many Power Companies came into being in a very short period of time in Europe and United States. Further developments were made in the 20 and the 21st century. The power systems were getting more efficient with passing time and have become a core area of electrical engineering field.

In technical terms, it is the mathematical product of voltage and current. Many devices use AC power while others use DC power. For example, digital equipment like computers use DC power and air conditioners, refrigerators and pumps use AC power.

Practical Uses

In many areas of the world, the governments maintain an electric network that can connect a number of electric generators by means of Power. This combination of Generators is called a "Power Grid". Users can buy the electricity directly from the Grid.

The basic component of Power transmission is a Transformer. Transformers can convert power from higher to lower voltages and viceversa. There are basically two types of transformers namely, Step-up and Step-down transformers. Electrical Substations are always present in a Power Grid to convert power to High voltages before transmission and to lower voltages after transmission.

Basically Power Engineering is a network of Inter-connected parts which convert different types of energy to electrical energy.

There are three basic subsystems of of Modern Power namely:

- Transmission subsystem.
- Generation subsystem.
- Distribution system.

These three systems produce the electricity, transmit it to load centers and transmit the power to the customers. Power generation means that the power is being transformed into electrical energy. Transmission system provides a means transporting electrical energy.

There are two types of transmission systems namely:

- Build transmission system.
- Sub-transmission system.

Power distribution system basically transports power from the transmission system to the user. As network systems are expensive so the distribution systems are normally radial. Voltage regulators, capacitors and relays are all components of the Control Equipment.

Application of Power Electronics in Electrical Engineering

Power Electronics is nothing but a way of converting electrical energy from one form to another, the output by the conversion is a better, more efficient, error free, clean, compact, simple and convenient to use.

The study of power electronics and electric motor drives involves processing of electric power for a variety of uses. And provide a wide range of application within the range from a few watts to mega-watts. For the example we can take various computer disk drives, drills with variable speed, lamp dimmers, electric vehicles, industrial machine tools, and electrical drives.

The area of power electronics is so vast that every day average 12 billion kilowatts per hour of power which is more than 80 % of the power generated is converted or processed or recycled by some of power electronic devices. As the power conversion efficiency is low the losses of energy during the power conversion are so high. We can understand this by a small example of estimated power consumption in a desktop sold in one year, which is equal to the 17 power plants of 500 MW. So it is very necessary to improve the efficiency of the power conversion systems.

By an estimated calculation, it has been found that with the use of efficient and economically effective power electronic technology, the world can save 35 % reduction in energy consumption.

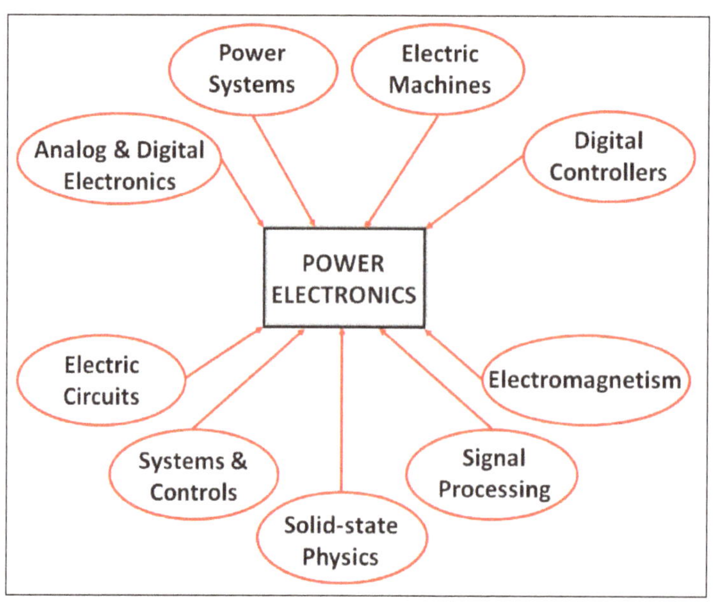

Following are various applications with a wide power range from a few tens of watts to several hundreds of megawatts. As power semiconductor devices improve in performance, efficiency and if the cost will be reduced, more systems will undoubtedly use power electronics.

Residential	
• Refrigeration and freezers • Space heating • Air conditioning	• Cooking • Lighting • Electronics (PCs, other entertainment equipment)
Commercial	
• Heating, ventilating, and air conditioning • Central refrigeration • Lighting • Computers and office equipment • Uninterruptible power supplies (UPSs) • Elevators • Industrial	• Pumps • Compressors • Blowers and fans • Machine tools (robots) • Arc furnaces and induction furnaces • Lighting • Industrial lasers • Induction heating • Welding
Transportation	
• Traction control of electric vehicles • Battery chargers for electric vehicles • Electric locomotives • Street cars, trolley buses • Subways • Automotive electronics, including engine controls • Utility systems • HVDC • SVC	• Supplemental energy sources (wind, photo voltaic), fuel cells • Energy storage systems • Induced draft fans and boiler feedwater pumps • Aerospace • Space shuttle power supply systems • Satellite power systems • Aircraft power systems • Telecommunications • Battery chargers • Power supplies (DC and UPS)

It is literally impossible to list all the application of power electronics today; it has covered almost all the areas where electrical energy is being used. This has become a trend now and it is increasing, especially with the present scenario of new devices and integrated design of power semiconductor devices and controllers. The ease of manufacturing has also helped these devices to be available so now a day these devices exist in a vast range of ratings and gradually have appeared in high voltage and extra high voltage systems also. In the end it can be summaries in the words that, the day is not far when all of the electrical energy in the world will pass through power electronic systems.

Control Engineering

Control engineering or control systems engineering is an engineering discipline that applies automatic control theory to design systems with desired behaviors in control environments. The discipline of controls overlaps and is usually taught along with electrical engineering at many institutions around the world.

The practice uses sensors and detectors to measure the output performance of the process being controlled; these measurements are used to provide corrective feedback helping to achieve the desired performance. Systems designed to perform without requiring human input are called automatic control systems (such as cruise control for regulating the speed of a car). Multi-disciplinary in nature, control systems engineering activities focus on implementation of control systems mainly derived by mathematical modeling of a diverse range of systems.

Modern day control engineering is a relatively new field of study that gained significant attention during the 20th century with the advancement of technology. It can be broadly defined or classified as practical application of control theory. Control engineering has an essential role in a wide range of control systems, from simple household washing machines to high-performance F-16 fighter aircraft. It seeks to understand physical systems, using mathematical modelling, in terms of inputs, outputs and various components with different behaviors; use control systems design tools to develop controllers for those systems; and implement controllers in physical systems employing available technology. A system can be mechanical, electrical, fluid, chemical, financial or biological, and the mathematical modelling, analysis and controller design uses control theory in one or many of the time, frequency and complex-s domains, depending on the nature of the design problem.

Control of fractionating columns is one of the more challenging applications.

Automatic control systems were first developed over two thousand years ago. The first feedback control device on record is thought to be the ancient Ktesibios's water clock in Alexandria, Egypt around the third century B.C.E. It kept time by regulating the water level in a vessel and, therefore, the water flow from that vessel. This certainly was a successful device as water clocks of similar design were still being made in Baghdad when the Mongols captured the city in 1258 A.D. A variety of automatic devices have been used over the centuries to accomplish useful tasks or simply just to entertain. The latter includes the automata, popular in Europe in the 17th and 18th centuries, featuring dancing figures that would repeat the same task over and over again; these automata are examples of open-loop control. Milestones among feedback, or "closed-loop" automatic control devices, include the temperature regulator of a furnace attributed to Drebbel, circa 1620, and the centrifugal flyball governor used for regulating the speed of steam engines by James Watt in 1788.

In his 1868 paper "On Governors", James Clerk Maxwell was able to explain instabilities exhibited by the flyball governor using differential equations to describe the control system. This demonstrated the importance and usefulness of mathematical models and methods in understanding complex phenomena, and it signaled the beginning of mathematical control and systems theory. Elements of control theory had appeared earlier but not as dramatically and convincingly as in Maxwell's analysis.

Control theory made significant strides over the next century. New mathematical techniques, as well as advancements in electronic and computer technologies, made it possible to control significantly more complex dynamical systems than the original flyball governor could stabilize. New mathematical techniques included developments in optimal control in the 1950s and 1960s followed by progress in stochastic, robust, adaptive, nonlinear, control methods in the 1970s and 1980s. Applications of control methodology have helped to make possible space travel and communication satellites, safer and more efficient aircraft, cleaner automobile engines, and cleaner and more efficient chemical processes.

Before it emerged as a unique discipline, control engineering was practiced as a part of mechanical engineering and control theory was studied as a part of electrical engineering since electrical circuits can often be easily described using control theory techniques. In the very first control relationships, a current output was represented by a voltage control input. However, not having adequate technology to implement electrical control systems, designers were left with the option of less efficient and slow responding mechanical systems. A very effective mechanical controller that is still widely used in some hydro plants is the governor. Later on, previous to modern power electronics, process control systems for industrial applications were devised by mechanical engineers using pneumatic and hydraulic control devices, many of which are still in use today.

Control Theory

There are two major divisions in control theory, namely, classical and modern, which have direct implications for the control engineering applications.

Classical SISO System Design

The scope of classical control theory is limited to single-input and single-output (SISO) system design, except when analyzing for disturbance rejection using a second input. The system analysis is carried out in the time domain using differential equations, in the complex-s domain with the Laplace transform, or in the frequency domain by transforming from the complex-s domain.

Many systems may be assumed to have a second order and single variable system response in the time domain. A controller designed using classical theory often requires on-site tuning due to incorrect design approximations. Yet, due to the easier physical implementation of classical controller designs as compared to systems designed using modern control theory, these controllers are preferred in most industrial applications. The most common controllers designed using classical control theory are PID controllers. A less common implementation may include either or both a Lead or Lag filter. The ultimate end goal is to meet requirements typically provided in the time-domain called the step response, or at times in the frequency domain called the open-loop response. The step response characteristics applied in a specification are typically percent overshoot, settling time, etc. The open-loop response characteristics applied in a specification are typically Gain and Phase margin and bandwidth. These characteristics may be evaluated through simulation including a dynamic model of the system under control coupled with the compensation model.

Modern MIMO System Design

Modern control theory is carried out in the state space, and can deal with multiple-input and multiple-output (MIMO) systems. This overcomes the limitations of classical control theory in more sophisticated design problems, such as fighter aircraft control, with the limitation that no frequency domain analysis is possible. In modern design, a system is represented to the greatest advantage as a set of decoupled first order differential equations defined using state variables. Nonlinear, multivariable, adaptive and robust control theories come under this division. Matrix methods are significantly limited for MIMO systems where linear independence cannot be assured in the relationship between inputs and outputs. Being fairly new, modern control theory has many areas yet to be explored. Scholars like Rudolf E. Kalman and Aleksandr Lyapunov are well-known among the people who have shaped modern control theory.

Control Systems

Control engineering is the engineering discipline that focuses on the modeling of a diverse range of dynamic systems (e.g. mechanical systems) and the design of controllers that will cause these systems to behave in the desired manner. Although such controllers need not be electrical, many are and hence control engineering is often viewed as a subfield of electrical engineering. However, the falling price of microprocessors is making the actual implementation of a control system essentially trivial. As a result, focus is shifting back to the mechanical and process engineering discipline, as intimate knowledge of the physical system being controlled is often desired.

Electrical circuits, digital signal processors and microcontrollers can all be used to implement control systems. Control engineering has a wide range of applications from the flight and propulsion systems of commercial airliners to the cruise control present in many modern automobiles.

In most cases, control engineers utilize feedback when designing control systems. This is often accomplished using a PID controller system. For example, in an automobile with cruise control the vehicle's speed is continuously monitored and fed back to the system, which adjusts the motor's torque accordingly. Where there is regular feedback, control theory can be used to determine how the system responds to such feedback. In practically all such systems stability is important and control theory can help ensure stability is achieved.

Although feedback is an important aspect of control engineering, control engineers may also work on the control of systems without feedback. This is known as open loop control. A classic example of open loop control is a washing machine that runs through a pre-determined cycle without the use of sensors.

Recent Advancement

Originally, control engineering was all about continuous systems. Development of computer control tools posed a requirement of discrete control system engineering because the communications between the computer-based digital controller and the physical system are governed by a computer clock. The equivalent to Laplace transform in the discrete domain is the Z-transform. Today, many of the control systems are computer controlled and they consist of both digital and analog components.

Therefore, at the design stage either digital components are mapped into the continuous domain and the design is carried out in the continuous domain, or analog components are mapped into discrete domain and design is carried out there. The first of these two methods is more commonly encountered in practice because many industrial systems have many continuous systems components, including mechanical, fluid, biological and analog electrical components, with a few digital controllers.

Similarly, the design technique has progressed from paper-and-ruler based manual design to computer-aided design and now to computer-automated design or CAutoD which has been made possible by evolutionary computation. CAutoD can be applied not just to tuning a predefined control scheme, but also to controller structure optimisation, system identification and invention of novel control systems, based purely upon a performance requirement, independent of any specific control scheme.

Resilient control systems extend the traditional focus of addressing only planned disturbances to frameworks and attempt to address multiple types of unexpected disturbance; in particular, adapting and transforming behaviors of the control system in response to malicious actors, abnormal failure modes, undesirable human action, etc.

Telecommunications Engineering

Telecommunications engineering is an engineering discipline centered on electrical and computer engineering which seeks to support and enhance telecommunication systems. The work ranges from basic circuit design to strategic mass developments. A telecommunication engineer is responsible for designing and overseeing the installation of telecommunications equipment and facilities, such as complex electronic switching systems, and other plain old telephone service facilities, optical fiber cabling, IP networks, and microwave transmission systems. Telecommunication engineering also overlaps with broadcast engineering.

Telecommunication is a diverse field of engineering connected to electronic, civil and systems engineering. they help find the cost of money for different types of computers and technological objects. Ultimately, telecom engineers are responsible for providing high-speed data transmission services. They use a variety of equipment and transport media to design the telecom network infrastructure; the most common media used by wired telecommunications today are twisted pair, coaxial cables, and optical fibers. Telecommunications engineers also provide solutions revolving around wireless modes of communication and information transfer, such as wireless telephony services, radio and satellite communications, and internet and broadband technologies.

Telecommunication systems are generally designed by telecommunication engineers which sprang from technological improvements in the telegraph industry in the late 19th century and the radio and the telephone industries in the early 20th century. Today, telecommunication is widespread and devices that assist the process, such as the television, radio and telephone, are common in many parts of the world. There are also many networks that connect these devices, including computer networks, public switched telephone network (PSTN), radio networks, and television networks. Computer communication across the Internet is one of many examples of telecommunication. Telecommunication plays a vital role in the world economy, and the telecommunication industry's revenue has been placed at just under 3% of the gross world product.

Telegraph and Telephone

Samuel Morse independently developed a version of the electrical telegraph that he unsuccessfully demonstrated on 2 September 1837. Soon after he was joined by Alfred Vail who developed the register — a telegraph terminal that integrated a logging device for recording messages to paper tape. This was demonstrated successfully over three miles (five kilometres) on 6 January 1838 and eventually over forty miles (sixty-four kilometres) between Washington, D.C. and Baltimore on 24 May 1844. The patented invention proved lucrative and by 1851 telegraph lines in the United States spanned over 20,000 miles (32,000 kilometres).

The first successful transatlantic telegraph cable was completed on 27 July 1866, allowing transatlantic telecommunication for the first time. Earlier transatlantic cables installed in 1857 and 1858 only operated for a few days or weeks before they failed. The international use of the telegraph has sometimes been dubbed the "Victorian Internet".

The first commercial telephone services were set up in 1878 and 1879 on both sides of the Atlantic in the cities of New Haven and London. Alexander Graham Bell held the master patent for the telephone that was needed for such services in both countries. The technology grew quickly from this point, with inter-city lines being built and telephone exchanges in every major city of the United States by the mid-1880s. Despite this, transatlantic voice communication remained impossible for customers until January 7, 1927 when a connection was established using radio. However no cable connection existed until TAT-1 was inaugurated on September 25, 1956 providing 36 telephone circuits.

In 1880, Bell and co-inventor Charles Sumner Tainter conducted the world's first wireless telephone call via modulated lightbeams projected by photophones. The scientific principles of their invention would not be utilized for several decades, when they were first deployed in military and fiber-optic communications.

Radio and Television

Marconi crystal radio receiver.

Over several years starting in 1894 the Italian inventor Guglielmo Marconi built the first complete, commercially successful wireless telegraphy system based on airborne electromagnetic waves (radio transmission). In December 1901, he would go on to established wireless communication between Britain and Newfoundland, earning him the Nobel Prize in physics in 1909 (which he shared with Karl Braun). In 1900 Reginald Fessenden was able to wirelessly transmit a human voice. On March 25, 1925, Scottish inventor John Logie Baird publicly demonstrated the transmission of moving silhouette pictures at the London department store Selfridges. In October 1925, Baird was successful in obtaining moving pictures with halftone shades, which were by most accounts the first true television pictures. This led to a public demonstration of the improved device on 26 January 1926 again at Selfridges. Baird's first devices relied upon the Nipkow disk and thus became known as the mechanical television. It formed the basis of semi-experimental broadcasts done by the British Broadcasting Corporation beginning September 30, 1929.

Satellite

The first U.S. satellite to relay communications was Project SCORE in 1958, which used a tape recorder to store and forward voice messages. It was used to send a Christmas greeting to the world from U.S. President Dwight D. Eisenhower. In 1960 NASA launched an Echo satellite; the 100-foot (30 m) aluminized PET film balloon served as a passive reflector for radio communications. Courier 1B, built by Philco, also launched in 1960, was the world's first active repeater satellite. Satellites these days are used for many applications such as uses in GPS, television, internet and telephone uses.

Telstar was the first active, direct relay commercial communications satellite. Belonging to AT&T as part of a multi-national agreement between AT&T, Bell Telephone Laboratories, NASA, the British General Post Office, and the French National PTT (Post Office) to develop satellite communications, it was launched by NASA from Cape Canaveral on July 10, 1962, the first privately sponsored space launch. Relay 1 was launched on December 13, 1962, and became the first satellite to broadcast across the Pacific on November 22, 1963.

The first and historically most important application for communication satellites was in intercontinental long distance telephony. The fixed Public Switched Telephone Network relays telephone calls from land line telephones to an earth station, where they are then transmitted a receiving satellite dish via a geostationary satellite in Earth orbit. Improvements in submarine communications cables, through the use of fiber-optics, caused some decline in the use of satellites for fixed telephony in the late 20th century, but they still exclusively service remote islands such as Ascension Island, Saint Helena, Diego Garcia, and Easter Island, where no submarine cables are in service. There are also some continents and some regions of countries where landline telecommunications are rare to nonexistent, for example Antarctica, plus large regions of Australia, South America, Africa, Northern Canada, China, Russia and Greenland.

After commercial long distance telephone service was established via communication satellites, a host of other commercial telecommunications were also adapted to similar satellites starting in 1979, including mobile satellite phones, satellite radio, satellite television and satellite Internet access. The earliest adaption for most such services occurred in the 1990s as the pricing for commercial satellite transponder channels continued to drop significantly.

Computer Networks and the Internet

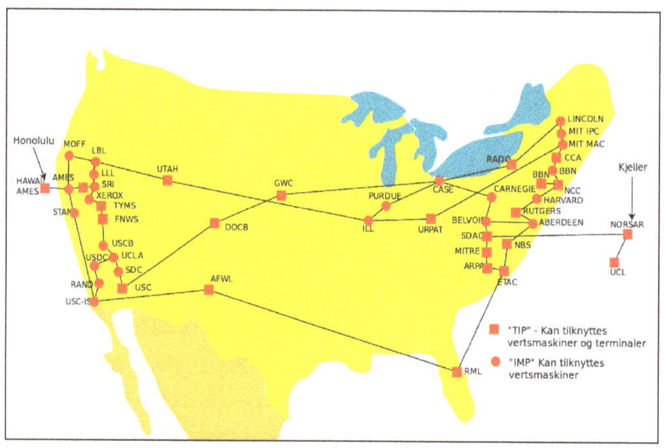

Symbolic representation of the Arpanet.

On 11 September 1940, George Stibitz was able to transmit problems using teleprinter to his Complex Number Calculator in New York and receive the computed results back at Dartmouth College in New Hampshire. This configuration of a centralized computer or mainframe computer with remote “dumb terminals” remained popular throughout the 1950s and into the 1960s. However, it was not until the 1960s that researchers started to investigate packet switching — a technology that allows chunks of data to be sent between different computers without first passing through a centralized mainframe. A four-node network emerged on 5 December 1969. This network soon became the ARPANET, which by 1981 would consist of 213 nodes.

ARPANET’s development centered around the Request for Comment process and on 7 April 1969, RFC 1 was published. This process is important because ARPANET would eventually merge with other networks to form the Internet, and many of the communication protocols that the Internet relies upon today were specified through the Request for Comment process. In September 1981, RFC 791 introduced the Internet Protocol version 4 (IPv4) and RFC 793 introduced the Transmission Control Protocol (TCP) — thus creating the TCP/IP protocol that much of the Internet relies upon today.

Optical Fiber

Optical fiber can be used as a medium for telecommunication and computer networking because it is flexible and can be bundled into cables. It is especially advantageous for long-distance communications, because light propagates through the fiber with little attenuation compared to electrical cables. This allows long distances to be spanned with few repeaters.

In 1966 Charles K. Kao and George Hockham proposed optical fibers at STC Laboratories (STL) at Harlow, England, when they showed that the losses of 1000 dB/km in existing glass (compared to 5-10 dB/km in coaxial cable) was due to contaminants, which could potentially be removed.

Optical fiber was successfully developed in 1970 by Corning Glass Works, with attenuation low enough for communication purposes (about 20dB/km), and at the same time GaAs (Gallium arsenide) semiconductor lasers were developed that were compact and therefore suitable for transmitting light through fiber optic cables for long distances.

After a period of research starting from 1975, the first commercial fiber-optic communications system was developed, which operated at a wavelength around 0.8 μm and used GaAs semiconductor lasers. This first-generation system operated at a bit rate of 45 Mbps with repeater spacing of up to 10 km. Soon on 22 April 1977, General Telephone and Electronics sent the first live telephone traffic through fiber optics at a 6 Mbit/s throughput in Long Beach, California.

The first wide area network fibre optic cable system in the world seems to have been installed by Rediffusion in Hastings, East Sussex, UK in 1978. The cables were placed in ducting throughout the town, and had over 1000 subscribers. They were used at that time for the transmission of television channels,not available because of local reception problems.

The first transatlantic telephone cable to use optical fiber was TAT-8, based on Desurvire optimized laser amplification technology. It went into operation in 1988.

In the late 1990s through 2000, industry promoters, and research companies such as KMI, and RHK predicted massive increases in demand for communications bandwidth due to increased use of the Internet, and commercialization of various bandwidth-intensive consumer services, such as video on demand. Internet protocol data traffic was increasing exponentially, at a faster rate than integrated circuit complexity had increased under Moore's Law.

Concepts

Radio transmitter room.

Basic Elements of a Telecommunication System

Transmitter

Transmitter (information source) that takes information and converts it to a signal for transmission. In electronics and telecommunications a transmitter or radio transmitter is an electronic device which, with the aid of an antenna, produces radio waves. In addition to their use in broadcasting, transmitters are necessary component parts of many electronic devices that communicate by radio, such as cell phones.

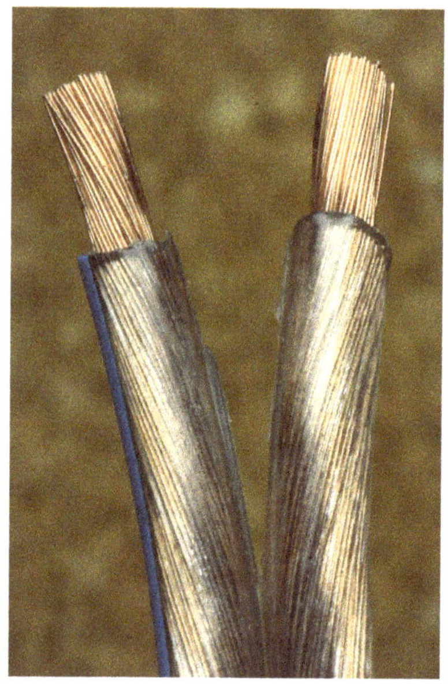

Copper wires.

Transmission Medium

Transmission medium over which the signal is transmitted. For example, the transmission medium for sounds is usually air, but solids and liquids may also act as transmission media for sound. Many transmission media are used as communications channel. One of the most common physical medias used in networking is copper wire. Copper wire is used to carry signals to long distances using relatively low amounts of power. Another example of a physical medium is optical fiber, which has emerged as the most commonly used transmission medium for long-distance communications. Optical fiber is a thin strand of glass that guides light along its length.

The absence of a material medium in vacuum may also constitute a transmission medium for electromagnetic waves such as light and radio waves.

Receiver

Receiver (information sink) that receives and converts the signal back into required information. In radio communications, a radio receiver is an electronic device that receives radio waves and converts the information carried by them to a usable form. It is used with an antenna. The information produced by the receiver may be in the form of sound (an audio signal), images (a video signal) or digital data.

Wireless communication tower, cell site.

Wired Communication

Wired communications make use of underground communications cables (less often, overhead lines), electronic signal amplifiers (repeaters) inserted into connecting cables at specified points, and terminal apparatus of various types, depending on the type of wired communications used.

Wireless Communication

Wireless communication involves the transmission of information over a distance without help of wires, cables or any other forms of electrical conductors. Wireless operations permit services, such as long-range communications, that are impossible or impractical to implement with the use of wires. The term is commonly used in the telecommunications industry to refer to telecommunications systems (e.g. radio transmitters and receivers, remote controls etc.) which use some form of energy (e.g. radio waves, acoustic energy, etc.) to transfer information without the use of wires. Information is transferred in this manner over both short and long distances.

Roles

Telecom Equipment Engineer

A telecom equipment engineer is an electronics engineer that designs equipment such as routers, switches, multiplexers, and other specialized computer/electronics equipment designed to be used in the telecommunication network infrastructure.

Network Engineer

A network engineer is a computer engineer who is in charge of designing, deploying and maintaining computer networks. In addition, they oversee network operations from a network operations center, designs backbone infrastructure, or supervises interconnections in a data center.

Central-office Engineer

Typical Northern Telecom DMS100 Telephone Central Office Installation.

A central-office engineer is responsible for designing and overseeing the implementation of telecommunications equipment in a central office (CO for short), also referred to as a wire center or telephone exchange A CO engineer is responsible for integrating new technology into the existing network, assigning the equipment's location in the wire center, and providing power, clocking (for digital equipment), and alarm monitoring facilities for the new equipment. The CO engineer is also responsible for providing more power, clocking, and alarm monitoring facilities if there are currently not enough available to support the new equipment being installed. Finally, the CO engineer is responsible for designing how the massive amounts of cable will be distributed to various equipment and wiring frames throughout the wire center and overseeing the installation and turn up of all new equipment.

Sub-roles

As structural engineers, CO engineers are responsible for the structural design and placement of racking and bays for the equipment to be installed in as well as for the plant to be placed on.

As electrical engineers, CO engineers are responsible for the resistance, capacitance, and inductance (RCL) design of all new plant to ensure telephone service is clear and crisp and data service is clean as well as reliable. Attenuation or gradual loss in intensity and loop loss calculations are required to determine cable length and size required to provide the service called for. In addition, power requirements have to be calculated and provided to power any electronic equipment being placed in the wire center.

Overall, CO engineers have seen new challenges emerging in the CO environment. With the advent of Data Centers, Internet Protocol (IP) facilities, cellular radio sites, and other emerging-technology equipment environments within telecommunication networks, it is important that a consistent set of established practices or requirements be implemented.

Installation suppliers or their sub-contractors are expected to provide requirements with their products, features, or services. These services might be associated with the installation of new or expanded equipment, as well as the removal of existing equipment.

Several other factors must be considered such as:

- Regulations and safety in installation.
- Removal of hazardous material.
- Commonly used tools to perform installation and removal of equipment.

Outside-plant Engineer

Engineers working on a cross-connect box, also known as a serving area interface.

Outside plant (OSP) engineers are also often called field engineers because they frequently spend much time in the field taking notes about the civil environment, aerial, above ground, and below ground. OSP engineers are responsible for taking plant (copper, fiber, etc.) from a wire center to a distribution point or destination point directly. If a distribution point design is used, then a cross-connect box is placed in a strategic location to feed a determined distribution area.

The cross-connect box, also known as a serving area interface, is then installed to allow connections to be made more easily from the wire center to the destination point and ties up fewer facilities by not having dedication facilities from the wire center to every destination point. The plant is then taken directly to its destination point or to another small closure called a terminal, where access can also be gained to the plant if necessary. These access points are preferred as they allow faster repair times for customers and save telephone operating companies large amounts of money.

The plant facilities can be delivered via underground facilities, either direct buried or through conduit or in some cases laid under water, via aerial facilities such as telephone or power poles, or via microwave radio signals for long distances where either of the other two methods is too costly.

Sub-roles

As structural engineers, OSP engineers are responsible for the structural design and placement of cellular towers and telephone poles as well as calculating pole capabilities of existing telephone or power poles onto which new plant is being added. Structural calculations are required when boring under heavy traffic areas such as highways or when attaching to other structures such as bridges. Shoring also has to be taken into consideration for larger trenches or pits. Conduit structures often include encasements of slurry that needs to be designed to support the structure and withstand the environment around it (soil type, high traffic areas, etc.).

Engineer (OSP) climbing a telephone pole.

As electrical engineers, OSP engineers are responsible for the resistance, capacitance, and inductance (RCL) design of all new plant to ensure telephone service is clear and crisp and data service is clean as well as reliable. Attenuation or gradual loss in intensity and loop loss calculations are required to determine cable length and size required to provide the service called for. In addition power requirements have to be calculated and provided to power any electronic equipment being placed in the field. Ground potential has to be taken into consideration when placing equipment, facilities, and plant in the field to account for lightning strikes, high voltage intercept from improperly grounded or broken power company facilities, and from various sources of electromagnetic interference.

As civil engineers, OSP engineers are responsible for drafting plans, either by hand or using Computer-aided design (CAD) software, for how telecom plant facilities will be placed. Often when working with municipalities trenching or boring permits are required and drawings must be made for these. Often these drawings include about 70% or so of the detailed information required to pave a road or add a turn lane to an existing street. Structural calculations are required when boring under heavy traffic areas such as highways or when attaching to other structures such as bridges. As civil engineers, telecom engineers provide the modern communications backbone for all technological communications distributed throughout civilizations today.

Unique to telecom engineering is the use of air-core cable which requires an extensive network of air handling equipment such as compressors, manifolds, regulators and hundreds of miles of air pipe per system that connects to pressurized splice cases all designed to pressurize this special form of copper cable to keep moisture out and provide a clean signal to the customer.

As political and social ambassador, the OSP engineer is a telephone operating company's face and voice to the local authorities and other utilities. OSP engineers often meet with municipalities, construction companies and other utility companies to address their concerns and educate them about how the telephone utility works and operates. Additionally, the OSP engineer has to secure real estate to place outside facilities on, such as an easement to place a cross-connect box on.

Electronic Engineering

Electronic engineering (also called electronics and communications engineering) is an electrical engineering discipline which utilizes nonlinear and active electrical components (such as semiconductor devices, especially transistors, diodes and integrated circuits) to design electronic circuits, devices, VLSI devices and their systems. The discipline typically also designs passive electrical components, usually based on printed circuit boards.

Electronics is a subfield within the wider electrical engineering academic subject but denotes a broad engineering field that covers subfields such as analog electronics, digital electronics, consumer electronics, embedded systems and power electronics. Electronics engineering deals with implementation of applications, principles and algorithms developed within many related fields, for example solid-state physics, radio engineering, telecommunications, control systems, signal processing, systems engineering, computer engineering, instrumentation engineering, electric power control, robotics, and many others.

The Institute of Electrical and Electronics Engineers (IEEE) is one of the most important and influential organizations for electronics engineers based in the US. On an international level, the International Electrotechnical Commission (IEC) prepares standards for electronic engineering, developed through consensus and thanks to the work of 20,000 experts from 172 countries worldwide.

Relationship to Electrical Engineering

Electronics is a subfield within the wider electrical engineering academic subject. An academic degree with a major in electronics engineering can be acquired from some universities, while other universities use electrical engineering as the subject. The term electrical engineer is still used in the academic world to include electronic engineers. However, some people consider the term 'electrical engineer' should be reserved for those having specialized in power and heavy current or high voltage engineering, while others consider that power is just one subset of electrical engineering, as well as 'electrical distribution engineering'. The term 'power engineering' is used as a descriptor in that industry. Again, in recent years there has been a growth of new separate-entry degree courses such as 'systems engineering' and 'communication systems engineering', often followed by academic departments of similar name, which are typically not considered as subfields of electronics engineering but of electrical engineering.

Electronics

In the field of electronic engineering, engineers design and test circuits that use the electromagnetic properties of electrical components such as resistors, capacitors, inductors, diodes and transistors to achieve a particular functionality. The tuner circuit, which allows the user of a radio to filter out all but a single station, is just one example of such a circuit.

In designing an integrated circuit, electronics engineers first construct circuit schematics that specify the electrical components and describe the interconnections between them. When completed, VLSI engineers convert the schematics into actual layouts, which map the layers of various conductor and semiconductor materials needed to construct the circuit. The conversion from schematics to layouts can be done by software but very often requires human fine-tuning to decrease space and power consumption. Once the layout is complete, it can be sent to a fabrication plant for manufacturing.

For systems of intermediate complexity, engineers may use VHDL modeling for programmable logic devices and FPGAs.

Integrated circuits, FPGAs and other electrical components can then be assembled on printed circuit boards to form more complicated circuits. Today, printed circuit boards are found in most electronic devices including televisions, computers and audio players.

Subfields

Electronic engineering has many subfields. This topic describes some of the most popular subfields in electronic engineering; although there are engineers who focus exclusively on one subfield, there are also many who focus on a combination of subfields.

Signal processing deals with the analysis and manipulation of signals. Signals can be either analog, in which case the signal varies continuously according to the information, or digital, in which case the signal varies according to a series of discrete values representing the information.

For analog signals, signal processing may involve the amplification and filtering of audio signals for audio equipment or the modulation and demodulation of signals for telecommunications. For digital signals, signal processing may involve the compression, error checking and error detection of digital signals.

Telecommunications engineering deals with the transmission of information across a channel such as a co-axial cable, optical fiber or free space.

Transmissions across free space require information to be encoded in a carrier wave in order to shift the information to a carrier frequency suitable for transmission, this is known as modulation. Popular analog modulation techniques include amplitude modulation and frequency modulation. The choice of modulation affects the cost and performance of a system and these two factors must be balanced carefully by the engineer.

Once the transmission characteristics of a system are determined, telecommunication engineers design the transmitters and receivers needed for such systems. These two are sometimes combined to form a two-way communication device known as a transceiver. A key consideration in the design of transmitters is their power consumption as this is closely related to their signal strength. If the signal strength of a transmitter is insufficient the signal's information will be corrupted by noise.

Electromagnetics is an in-depth study about the signals that are transmitted in a channel (Wired or Wireless). This includes Basics of Electromagnetic waves, Transmission Lines and Waveguides, Antennas, its types and applications with Radio-Frequency (RF) and Microwaves. Its applications are seen widely in other sub-fields like Telecommunication, Control and Instrumentation Engineering.

Control engineering has a wide range of applications from the flight and propulsion systems of commercial airplanes to the cruise control present in many modern cars. It also plays an important role in industrial automation.

Control engineers often utilize feedback when designing control systems. For example, in a car with cruise control, the vehicle's speed is continuously monitored and fed back to the system which adjusts the engine's power output accordingly. Where there is regular feedback, control theory can be used to determine how the system responds to such feedback.

Instrumentation engineering deals with the design of devices to measure physical quantities such as pressure, flow and temperature. These devices are known as instrumentation.

The design of such instrumentation requires a good understanding of physics that often extends beyond electromagnetic theory. For example, radar guns use the Doppler effect to measure the speed of oncoming vehicles. Similarly, thermocouples use the Peltier–Seebeck effect to measure the temperature difference between two points.

Often instrumentation is not used by itself, but instead as the sensors of larger electrical systems. For example, a thermocouple might be used to help ensure a furnace's temperature remains constant. For this reason, instrumentation engineering is often viewed as the counterpart of control engineering.

Computer engineering deals with the design of computers and computer systems. This may involve the design of new computer hardware, the design of PDAs or the use of computers to control an industrial plant. Development of embedded systems—systems made for specific tasks (e.g., mobile phones)—is also included in this field. This field includes the micro controller and its applications. Computer engineers may also work on a system's software. However, the design of complex software systems is often the domain of software engineering, which is usually considered a separate discipline.

VLSI design engineering VLSI stands for *very large scale integration*. It deals with fabrication of ICs and various electronic components.

Signal Processing

Signal processing is an electrical engineering subfield that focuses on analysing, modifying and synthesizing signals such as sound, images and biological measurements. Signal processing techniques can be used to improve transmission, storage efficiency and subjective quality and to also emphasize or detect components of interest in a measured signal.

According to Alan V. Oppenheim and Ronald W. Schafer, the principles of signal processing can be found in the classical numerical analysis techniques of the 17th century. Oppenheim and Schafer further state that the digital refinement of these techniques can be found in the digital control systems of the 1940s and 1950s.

In 1948, Claude Shannon wrote the influential paper "A Mathematical Theory of Communication" which was published in the Bell System Technical Journal. The paper laid the groundwork for later development of information communication systems. Around the same time, methods of signal transmission were being rapidly developed, as a new type of signal emerged called *processing signals*. Single-chip digital signal processing was introduced in 1979.

Categories

Analog

Analog signal processing is for signals that have not been digitized, as in legacy radio, telephone, radar, and television systems. This involves linear electronic circuits as well as non-linear ones. The former are, for instance, passive filters, active filters, additive mixers, integrators and delay lines. Non-linear circuits include compandors, multiplicators (frequency mixers and voltage-controlled amplifiers), voltage-controlled filters, voltage-controlled oscillators and phase-locked loops.

Continuous Time

Continuous-time signal processing is for signals that vary with the change of continuous domain (without considering some individual interrupted points).

The methods of signal processing include time domain, frequency domain, and complex frequency domain. This technology mainly discusses the modeling of linear time-invariant continuous system, integral of the system's zero-state response, setting up system function and the continuous time filtering of deterministic signals.

Discrete Time

Discrete-time signal processing is for sampled signals, defined only at discrete points in time, and as such are quantized in time, but not in magnitude.

Analog discrete-time signal processing is a technology based on electronic devices such as sample and hold circuits, analog time-division multiplexers, analog delay lines and analog feedback shift registers. This technology was a predecessor of digital signal processing, and is still used in advanced processing of gigahertz signals.

The concept of discrete-time signal processing also refers to a theoretical discipline that establishes a mathematical basis for digital signal processing, without taking quantization error into consideration.

Digital

Digital signal processing is the processing of digitized discrete-time sampled signals. Processing is done by general-purpose computers or by digital circuits such as ASICs, field-programmable gate arrays or specialized digital signal processors (DSP chips). Typical arithmetical operations include fixed-point and floating-point, real-valued and complex-valued, multiplication and addition. Other typical operations supported by the hardware are circular buffers and lookup tables. Examples of algorithms are the Fast Fourier transform (FFT), finite impulse response (FIR) filter, Infinite impulse response (IIR) filter, and adaptive filters such as the Wiener and Kalman filters.

Nonlinear

Nonlinear signal processing involves the analysis and processing of signals produced from nonlinear systems and can be in the time, frequency, or spatio-temporal domains. Nonlinear systems can produce highly complex behaviors including bifurcations, chaos, harmonics, and subharmonics which cannot be produced or analyzed using linear methods.

Statistical

Statistical signal processing is an approach which treats signals as stochastic processes, utilizing their statistical properties to perform signal processing tasks. Statistical techniques are widely used in signal processing applications. For example, one can model the probability distribution of noise incurred when photographing an image, and construct techniques based on this model to reduce the noise in the resulting image.

Application Fields

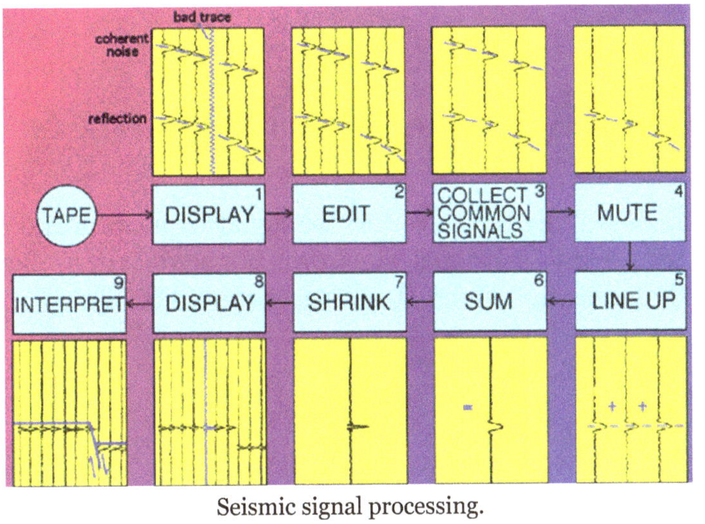

Seismic signal processing.

- Audio signal processing – for electrical signals representing sound, such as speech or music.
- Speech signal processing – for processing and interpreting spoken words.
- Image processing – in digital cameras, computers and various imaging systems.
- Video processing – for interpreting moving pictures.
- Wireless communication – waveform generations, demodulation, filtering, equalization.
- Control systems.
- Array processing – for processing signals from arrays of sensors.
- Process control – a variety of signals are used, including the industry standard 4-20 mA current loop.
- Seismology.
- Financial signal processing – analyzing financial data using signal processing techniques, especially for prediction purposes.
- Feature extraction, such as image understanding and speech recognition.

- Quality improvement, such as noise reduction, image enhancement, and echo cancellation.
- (Source coding), including audio compression, image compression, and video compression.
- Genomics, Genomic signal processing.

In communication systems, signal processing may occur at:

- OSI layer 1 in the seven layer OSI model, the Physical Layer (modulation, equalization, multiplexing, etc.).
- OSI layer 2, the Data Link Layer (Forward Error Correction).
- OSI layer 6, the Presentation Layer (source coding, including analog-to-digital conversion and signal compression).

Typical Devices

- Filters – for example analog (passive or active) or digital (FIR, IIR, frequency domain or stochastic filters, etc.)
- Samplers and analog-to-digital converters for signal acquisition and reconstruction, which involves measuring a physical signal, storing or transferring it as digital signal, and possibly later rebuilding the original signal or an approximation thereof.
- Signal compressors.
- Digital signal processors (DSPs).

Instrumentation

Instrumentation is the branch of engineering that deals with measurement and control.

According to ISA, the official definition of Instrumentation – is a collection ofInstruments and their application for the purpose of Observation, Measurement and Control.

Instrumentation and Control

An instrument is a device that measures or manipulates process physical variables such as flow, temperature, level, or pressure etc. Instruments include many varied contrivances which can be as simple as valves and transmitters, and as complex as analyzers. Instruments often comprise control systems of varied processes. The control of processes is one of the main branches of applied instrumentation.

Control instrumentation includes devices such as solenoids, valves, circuit breakers, and relays. These devices are able to change a field parameter, and provide remote or automated control capabilities.

Transmitters are devices which produce an analog signal, usually in the form of a 4–20 ma electrical current signal, although many other options using voltage,frequency, or pressure are possible.

This signal can be used to control other instruments directly, or it can be sent to a PLC, DCS, SCADA system, or other type of computerized controller, where it can be interpreted into readable values and used to control other devices and processes in the system.

Instrumentation plays a significant role in both gathering information from the field and changing the field parameters, and as such are a key part of control loops.

Measurement

Instrumentation can be used to measure certain field parameters (physical values).

These measured values may include:

- Pressure,
- Either differential or static,
- Flow,
- Temperature,
- Level,
- Density,
- Viscosity,
- Radiation,
- Process instrumentation etc.

Instrumentation Engineering

Instrumentation engineering is the engineering specialization focused on the principle and operation of measuring instruments which are used in design and configuration of automated systems in electrical, pneumatic domains etc.

They typically work for industries with automated processes, such as chemical or manufacturing plants, with the goal of improving system productivity, reliability, safety, optimization and stability.

To control the parameters in a process or in a particular system, devices such as microprocessors, microcontrollers or PLCs are used, but their ultimate aim is to control the parameters of a system.

References

- Billings, S. A. (2013). Nonlinear System Identification: NARMAX Methods in the Time, Frequency, and Spatio-Temporal Domains. Wiley. ISBN 1119943590
- 52283-power-engineering-basics-and-job-profiles-explained, power-generation-distribution: brighthubengineering.com, Retrieved 16 January, 2019

- "Systems & Control Engineering FAQ | Electrical Engineering and Computer Science". Engineering.case.edu. Case Western Reserve University. 20 November 2015. Retrieved 27 June 2017
- Application-of-power-electronics-in-electrical-engineering, our-blog: aryacollege.org, Retrieved 29 March, 2019
- Klooster, John W. (2009). Icons of invention: the makers of the modern world from Gutenberg to Gates. ABC-CLIO. Pp. 161–168. ISBN 9780313347436. Retrieved 22 June 2017
- What-is-instrumentation-and-control-engineering: instrumentationtools.com, Retrieved 2 June, 2019
- Charles A. Harper High Performance Printed Circuit Boards, pp. Xiii-xiv, mcgraw-Hill Professional, 2000 ISBN 978-0-07-026713-8

3 Electrical Circuits

The closed path through which the electrical current flows is known as an electrical circuit. Some of the different types of electrical circuits are series circuits, parallel circuits, AC circuits and DC circuits. The chapter closely examines these types of electrical circuits to provide an extensive understanding of the subject.

Electric circuit is a path for transmitting electric current. An electric circuit includes a device that gives energy to the charged particles constituting the current, such as a battery or a generator; devices that use current, such as lamps, electric motors, or computers; and the connecting wires or transmission lines. Two of the basic laws that mathematically describe the performance of electric circuits are Ohm's law and Kirchhoff's rules.

Electric circuits are classified in several ways. A direct-current circuit carries current that flows only in one direction. An alternating-current circuit carries current that pulsates back and forth many times each second, as in most household circuits. A series circuit comprises a path along which the whole current flows through each component. A parallel circuit comprises branches so that the current divides and only part of it flows through any branch. The voltage, or potential difference, across each branch of a parallel circuit is the same, but the currents may vary. In a home electrical circuit, for instance, the same voltage is applied across each light or appliance, but each of these loads draws a different amount of current, according to its power requirements. A number of similar batteries connected in parallel provides greater current than a single battery, but the voltage is the same as for a single battery.

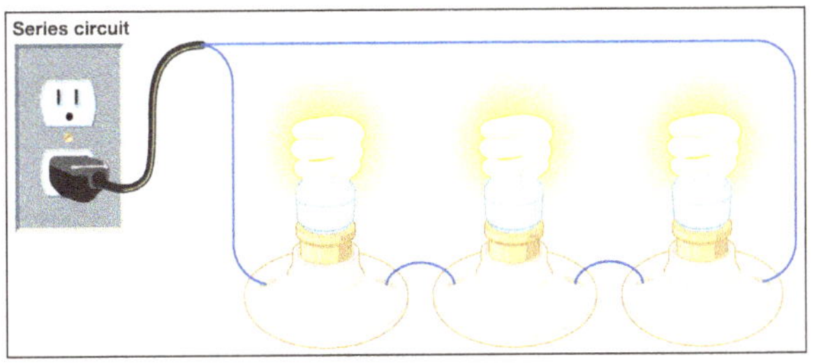

A series circuit.

The network of transistors, transformers, capacitors, connecting wires, and other electroniccomponents within a single device such as a radio is also an electric circuit. Such complex circuits may be made up of one or more branches in combinations of series and series-parallel arrangements.

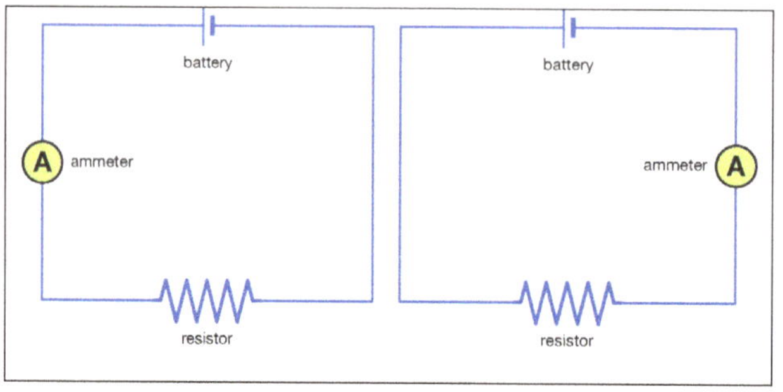

Ammeter: Two diagrams showing an ammeter connected to a simple circuit in two different positions.

Series and Parallel Circuits

Components of an electrical circuit or electronic circuit can be connected in series, parallel, or series-parallel. The two simplest of these are called series and parallel and occur frequently. Components connected in series are connected along a single conductive path, so the same current flows through all of the components but voltage is dropped (lost) across each of the resistances. In a series circuit, the sum of the voltages consumed by each individual resistance is equal to the source voltage. Components connected in parallel are connected along multiple paths so that the current can split up; the same voltage is applied to each component.

A circuit composed solely of components connected in series is known as a series circuit; likewise, one connected completely in parallel is known as a parallel circuit. In a series circuit, the current that flows through each of the components is the same, and the voltage across the circuit is the sum of the individual voltage drops across each component. In a parallel circuit, the voltage across each of the components is the same, and the total current is the sum of the currents flowing through each component.

Consider a very simple circuit consisting of four light bulbs and a 12-volt automotive battery. If a wire joins the battery to one bulb, to the next bulb, to the next bulb, to the next bulb, then back to the battery in one continuous loop, the bulbs are said to be in series. If each bulb is wired to the battery in a separate loop, the bulbs are said to be in parallel. If the four light bulbs are connected in series, the same amperage flows through all of them and the voltage drop is 3-volts across each bulb, which may not be sufficient to make them glow. If the light bulbs are connected in parallel, the currents through the light bulbs combine to form the current in the battery, while the voltage drop is 12-volts across each bulb and they all glow.

In a series circuit, every device must function for the circuit to be complete. If one bulb burns out in a series circuit, the entire circuit is broken. In parallel circuits, each light bulb has its own circuit, so all but one light could be burned out, and the last one will still function.

Series Circuits

Series circuits are sometimes referred to as *current*-coupled or daisy chain-coupled. The current in a series circuit goes through every component in the circuit. Therefore, all of the components in a series connection carry the same current.

A series circuit has only one path in which its current can flow. Opening or breaking a series circuit at any point causes the entire circuit to "open" or stop operating. For example, if even one of the light bulbs in an older-style string of Christmas tree lights burns out or is removed, the entire string becomes inoperable until the bulb is replaced.

Current

$$I = I_1 = I_2 = \cdots = I_n$$

In a series circuit, the current is the same for all of the elements.

Voltage

In a series circuit, the voltage is the sum of the voltage drops of the individual components (resistance units).

$$V = V_1 + V_2 + \ldots + V_n$$

Resistance Units

The total resistance of resistance units in series is equal to the sum of their individual resistances:

$$R_{\text{total}} = R_s = R_1 + R_2 + \cdots + R_n$$

Rs=>Resistance in series

Electrical conductance presents a reciprocal quantity to resistance. Total conductance of a series circuits of pure resistances, therefore, can be calculated from the following expression:

$$\frac{1}{G_{\text{total}}} = \frac{1}{G_1} + \frac{1}{G_2} + \cdots + \frac{1}{G_n}.$$

For a special case of two resistances in series, the total conductance is equal to:

$$G_{\text{total}} = \frac{G_1 G_2}{G_1 + G_2}.$$

Inductors

Inductors follow the same law, in that the total inductance of non-coupled inductors in series is equal to the sum of their individual inductances:

$$L_{\text{total}} = L_1 + L_2 + \cdots + L_n$$

However, in some situations, it is difficult to prevent adjacent inductors from influencing each other, as the magnetic field of one device coupled with the windings of its neighbours. This influence is defined by the mutual inductance M. For example if two inductors are in series, there are two possible equivalent inductances depending on how the magnetic fields of both inductors influence each other.

When there are more than two inductors, the mutual inductance between each of them and the way the coils influence each other complicates the calculation. For a larger number of coils the total combined inductance is given by the sum of all mutual inductances between the various coils including the mutual inductance of each given coil with itself, which we term self-inductance or simply inductance. For three coils, there are six mutual inductances M_{12}, M_{13}, M_{23} and M_{21}, M_{31} and M_{32}. There are also the three self-inductances of the three coils: M_{11}, M_{22} and M_{33}.

Therefore,

$$L_{\text{total}} = (M_{11} + M_{22} + M_{33}) + (M_{12} + M_{13} + M_{23}) + (M_{21} + M_{31} + M_{32})$$

By reciprocity $M_{ij} = M_{ji}$ so that the last two groups can be combined. The first three terms represent the sum of the self-inductances of the various coils. The formula is easily extended to any number of series coils with mutual coupling. The method can be used to find the self-inductance of large coils of wire of any cross-sectional shape by computing the sum of the mutual inductance of each turn of wire in the coil with every other turn since in such a coil all turns are in series.

Capacitors

Capacitors follow the same law using the reciprocals. The total capacitance of capacitors in series is equal to the reciprocal of the sum of the reciprocals of their individual capacitances:

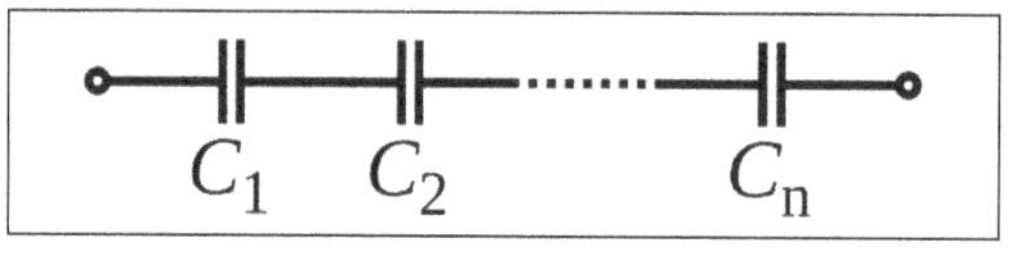

$$\frac{1}{C_{\text{total}}} = \frac{1}{C_1} + \frac{1}{C_2} + \cdots + \frac{1}{C_n}.$$

Switches

Two or more switches in series form a logical AND; the circuit only carries current if all switches are closed.

Cells and Batteries

A battery is a collection of electrochemical cells. If the cells are connected in series, the voltage of the battery will be the sum of the cell voltages. For example, a 12 volt car battery contains six 2-volt cells connected in series. Some vehicles, such as trucks, have two 12 volt batteries in series to feed the 24-volt system.

Parallel Circuits

If two or more components are connected in parallel, they have the same difference of potential (voltage) across their ends. The potential differences across the components are the same in magnitude, and they also have identical polarities. The same voltage is applied to all circuit components connected in parallel. The total current is the sum of the currents through the individual components, in accordance with Kirchhoff's current law.

Voltage

In a parallel circuit, the voltage is the same for all elements.

$$V = V_1 = V_2 = \ldots = V_n$$

Current

The current in each individual resistor is found by Ohm's law. Factoring out the voltage gives

$$I_{\text{total}} = V\left(\frac{1}{R_1} + \frac{1}{R_2} + \cdots + \frac{1}{R_n}\right).$$

Resistance Units

To find the total resistance of all components, add the reciprocals of the resistances R_i of each component and take the reciprocal of the sum. Total resistance will always be less than the value of the smallest resistance:

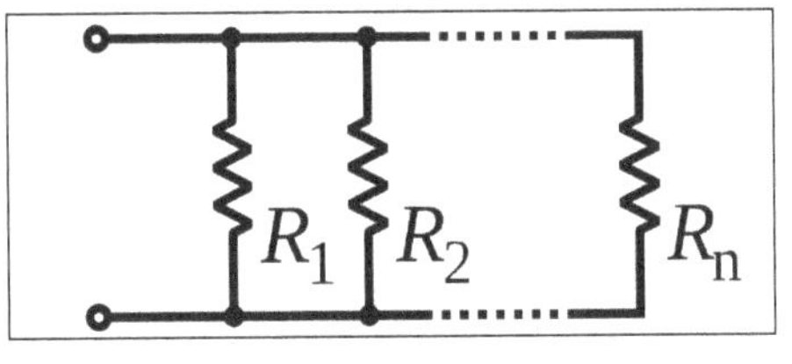

$$\frac{1}{R_{\text{total}}} = \frac{1}{R_1} + \frac{1}{R_2} + \cdots + \frac{1}{R_n}.$$

For only two resistances, the unreciprocated expression is reasonably simple:

$$R_{\text{total}} = \frac{R_1 R_2}{R_1 + R_2}.$$

This sometimes goes by the mnemonic *product over sum.*

For N equal resistances in parallel, the reciprocal sum expression simplifies to:

$$\frac{1}{R_{\text{total}}} = N\frac{1}{R}.$$

and therefore to:

$$R_{\text{total}} = \frac{R}{N}.$$

To find the current in a component with resistance R_i, use Ohm's law again:

$$I_i = \frac{V}{R_i}.$$

The components divide the current according to their reciprocal resistances, so, in the case of two resistors,

$$\frac{I_1}{I_2} = \frac{R_2}{R_1}.$$

An old term for devices connected in parallel is *multiple*, such as multiple connections for arc lamps.

Since electrical conductance G is reciprocal to resistance, the expression for total conductance of a parallel circuit of resistors reads:

$$G_{\text{total}} = G_1 + G_2 + \cdots + G_n.$$

The relations for total conductance and resistance stand in a complementary relationship: the expression for a series connection of resistances is the same as for parallel connection of conductances, and vice versa.

Inductors

Inductors follow the same law, in that the total inductance of non-coupled inductors in parallel is equal to the reciprocal of the sum of the reciprocals of their individual inductances:

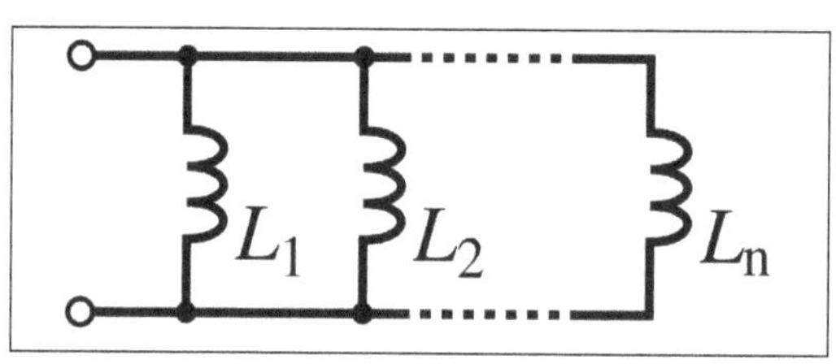

$$\frac{1}{L_{\text{total}}} = \frac{1}{L_1} + \frac{1}{L_2} + \cdots + \frac{1}{L_n}.$$

If the inductors are situated in each other's magnetic fields, this approach is invalid due to mutual inductance. If the mutual inductance between two coils in parallel is M, the equivalent inductor is:

$$\frac{1}{L_{\text{total}}} = \frac{L_1 + L_2 - 2M}{L_1 L_2 - M^2}$$

If, $L_1 = L_2$

$$L_{\text{total}} = \frac{L+M}{2}$$

The sign of M depends on how the magnetic fields influence each other. For two equal tightly coupled coils the total inductance is close to that of every single coil. If the polarity of one coil is reversed so that M is negative, then the parallel inductance is nearly zero or the combination is almost non-inductive. It is assumed in the "tightly coupled" case M is very nearly equal to L. However if the inductances are not equal and the coils are tightly coupled there can be near short circuit conditions and high circulating currents for both positive and negative values of M, which can cause problems.

More than three inductors become more complex and the mutual inductance of each inductor on each other inductor and their influence on each other must be considered. For three coils, there are three mutual inductances M_{12}, M_{13} and M_{23}. This is best handled by matrix methods and summing the terms of the inverse of the L matrix (3 by 3 in this case).

The pertinent equations are of the form: $v_i = \sum_j L_{i,j} \frac{di_j}{dt}$.

Capacitors

The total capacitance of capacitors in parallel is equal to the sum of their individual capacitances:

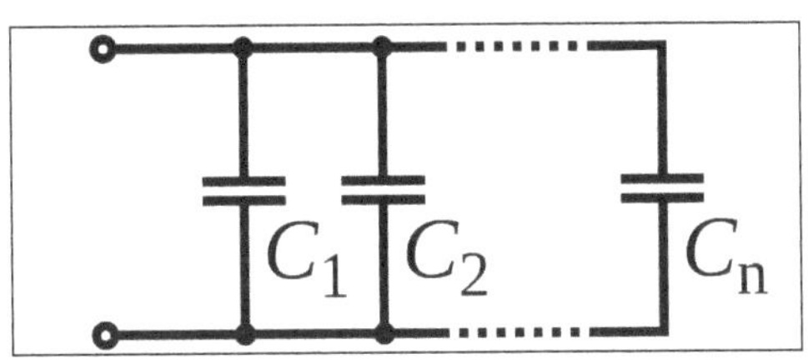

$$C_{\text{total}} = C_1 + C_2 + \cdots + C_n .$$

The working voltage of a parallel combination of capacitors is always limited by the smallest working voltage of an individual capacitor.

Switches

Two or more switches in parallel form a logical OR; the circuit carries current if at least one switch is closed.

Cells and Batteries

If the cells of a battery are connected in parallel, the battery voltage will be the same as the cell voltage, but the current supplied by each cell will be a fraction of the total current. For example, if a battery comprises four identical cells connected in parallel and delivers a current of 1 ampere, the current supplied by each cell will be 0.25 ampere. Parallel-connected batteries were widely used to power the valve filaments in portable radios. Lithium-ion rechargeable batteries (particularly laptop batteries) are often connected in parallel to increase the ampere-hour rating. Some solar electric systems have batteries in parallel to increase the storage capacity; a close approximation of total amp-hours is the sum of all amp-hours of in-parallel batteries.

Combining Conductances

From Kirchhoff's circuit laws we can deduce the rules for combining conductances. For two conductances G_1 and G_2 in *parallel*, the voltage across them is the same and from Kirchhoff's current law (KCL) the total current is,

$$I_{eq} = I_1 + I_2.$$

Substituting Ohm's law for conductances gives,

$$G_{eq}V = G_1V + G_2V$$

and the equivalent conductance will be,

$$G_{eq} = G_1 + G_2.$$

For two conductances and G_2 in series the current through them will be the same and Kirchhoff's Voltage Law tells us that the voltage across them is the sum of the voltages across each conductance, that is,

$$V_{eq} = V_1 + V_2.$$

Substituting Ohm's law for conductance then gives,

$$\frac{I}{G_{eq}} = \frac{I}{G_1} + \frac{I}{G_2}$$

which in turn gives the formula for the equivalent conductance,

$$\frac{1}{G_{eq}} = \frac{1}{G_1} + \frac{1}{G_2}.$$

This equation can be rearranged slightly, though this is a special case that will only rearrange like this for two components.

$$G_{eq} = \frac{G_1G_2}{G_1 + G_2}.$$

Notation

The value of two components in parallel is often represented in equations by the parallel operator, two vertical lines (||), borrowing the parallel lines notation from geometry.

$$R_{eq} \equiv R_1 \parallel R_2 \equiv \left(R_1^{-1} + R_2^{-1}\right)^{-1} \equiv \frac{R_1R_2}{R_1 + R_2}$$

This simplifies expressions that would otherwise become complicated by expansion of the terms. For instance:

$$R_1 \| R_2 \| R_3 \equiv \frac{R_1 R_2 R_3}{R_1 R_2 + R_1 R_3 + R_2 R_3}.$$

Applications

A common application of series circuit in consumer electronics is in batteries, where several cells connected in series are used to obtain a convenient operating voltage. Two disposable zinc cells in series might power a flashlight or remote control at 3 volts; the battery pack for a hand-held power tool might contain a dozen lithium-ion cells wired in series to provide 48 volts.

Series circuits were formerly used for lighting in electric multiple units trains. For example, if the supply voltage was 600 volts there might be eight 70-volt bulbs in series (total 560 volts) plus a resistor to drop the remaining 40 volts. Series circuits for train lighting were superseded, first by motor-generators, then by solid state devices.

Series resistance can also be applied to the arrangement of blood vessels within a given organ. Each organ is supplied by a large artery, smaller arteries, arterioles, capillaries, and veins arranged in series. The total resistance is the sum of the individual resistances, as expressed by the following equation: $R_{total} = R_{artery} + R_{arterioles} + R_{capillaries}$. The largest proportion of resistance in this series is contributed by the arterioles.

Parallel resistance is illustrated by the circulatory system. Each organ is supplied by an artery that branches off the aorta. The total resistance of this parallel arrangement is expressed by the following equation: $1/R_{total} = 1/R_a + 1/R_b + ... 1/R_n$. R_a, R_b, and R_n are the resistances of the renal, hepatic, and other arteries respectively. The total resistance is less than the resistance of any of the individual arteries.

AC Circuit

The path for the flow of alternating current is called an AC Circuit. The alternating current (AC) is used for domestic and industrial purposes. In an AC circuit, the value of the magnitude and the direction of current and voltages is not constant, it changes at a regular interval of time. It travels as a sinusoidal wave completing one cycle as half positive and half negative cycle and is a function of time (t) or angle (θ=wt).

In DC Circuit, the opposition to the flow of current is the only resistance of the circuit whereas the opposition to the flow of current in the AC circuit is because of resistance (R), Inductive Reactance ($X_L = 2\pi fL$) and capacitive reactance ($X_C = 1/2\ \pi fC$) of the circuit.

In AC Circuit, the current and voltages are represented by magnitude and direction. The alternating quantity may or may not be in phase with each other depending upon the various parameters of the circuit like resistance, inductance, and capacitance. The sinusoidal alternating quantities are voltage and current which varies according to the sine of angle θ.

For the generation of electric power, in all over the world the sinusoidal voltage and current are selected because of the following reasons are given below.

- The sinusoidal voltage and current produce low iron and copper losses in the transformer and rotating electrical machines, which in turns improves the efficiency of the AC machines.
- They offer less interference to the nearby communication system.
- They produce less disturbance in the electrical circuit.

Alternating Voltage and Current in an AC Circuit

The voltage that changes its polarity and magnitude at regular interval of time is called an alternating voltage. Similarly the direction of the current is changed and the magnitude of current changes with time it is called alternating current.When an alternating voltage source is connected across a load resistance as shown in the figure below, the current through it flows in one direction and then in the opposite direction when the polarity is reversed.

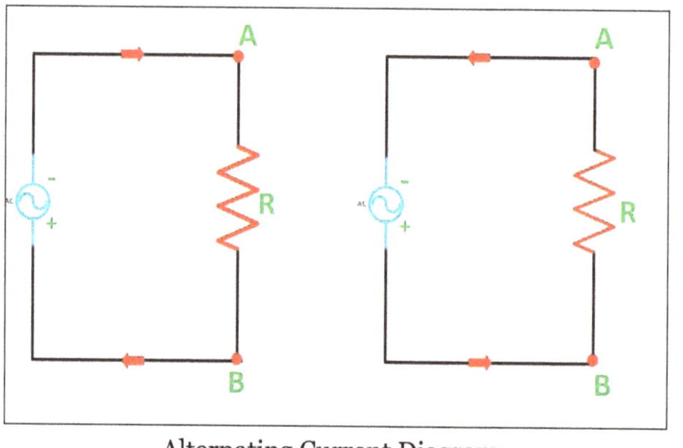

Alternating Current Diagram.

The waveform of the alternating voltage with respect to the time and the current flowing through the resistance (R) in the circuit is shown below.

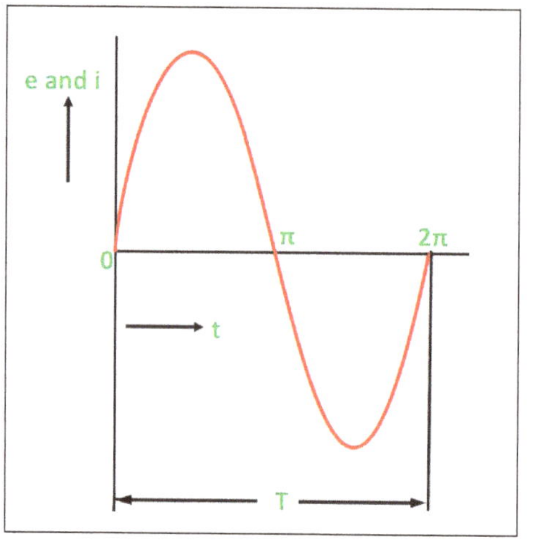

There are various types of AC circuit such as AC circuit containing only resistance (R), AC circuit containing only capacitance (C), AC circuit containing only inductance (L), the combination of RL

Circuit, AC circuit containing resistance and capacitance (RC), AC circuit containing inductance and capacitance (LC) and resistance inductance and capacitance (RLC) AC circuit.

The various terms which are frequently used in an AC Circuit are as follows:

- Amplitude: The maximum positive or negative value attained by an alternating quantity in one complete cycle is called Amplitude or peak value or maximum value. The maximum value of voltage and current is represented by E_m or V_m and I_m respectively.
- Alternation: One half cycle is termed as alternation. An alternation span is of 180 degrees electrical.
- Cycle: When one set of positive and negative values completes by an alternating quantity or it goes through 360 degrees electrical, it is said to have one complete Cycle.
- Instantaneous Value: The value of voltage or current at any instant of time is called an instantaneous value.It is denoted by (i or e).
- Frequency: The number of cycles made per second by an alternating quantity is called frequency. It is measured in cycle per second (c/s) or hertz (Hz) and is denoted by (f).
- Time Period: The time taken in seconds by a voltage or a current to complete one cycle is called Time Period. It is denoted by (T).
- Wave Form: The shape obtained by plotting the instantaneous values of an alternating quantity such as voltage and current along the y axis and the time (t) or angle (θ=wt) along the x axis is called waveform.

DC Circuit

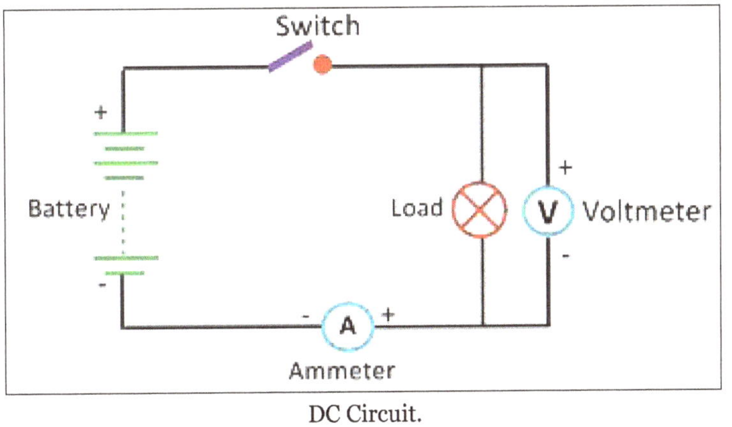

DC Circuit.

The closed path in which the direct current flows is called the DC circuit. The current flows in only one direction and it is mostly used in low voltage applications. The resistor is the main component of the DC circuit.

A simple DC circuit is shown in the figure below which contains a DC source (battery), a load lamp, a switch, connecting leads, and measuring instruments like ammeter and voltmeter. The load resistor is connected in series, parallel or series-parallel combination as per requirement.

Types of DC Circuit

The DC electric circuit is mainly classified into three groups. They are the series DC circuit, parallel DC circuit, and series and parallel DC circuit.

DC Series Circuit

The circuit which have DC series source, and the number of resistors are connected end to end so that same current flow through them is called a DC series circuit. The figure below shows the simple series circuit. In the series circuit the resistor R_1, R_2, and R_3 are connected in series across a supply voltage of V volts. The same current I is flowing through all the three resistors.

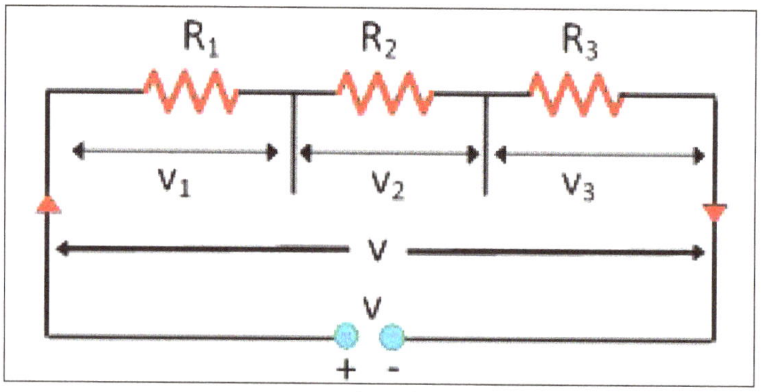

DC Series Circuit.

If V_1, V_2, and V_3 are the voltage drop across the three resistor R_1, R_2, and R_3 respectively, then

$$V = V_1 + V_2 + V_3$$
$$V = IR_1 + IR_2 + IR_3$$

Let R be the total resistance of the circuit then,

$$IR = IR_1 + IR_2 + IR_3$$
$$R = R_1 + R_2 + R_3$$

Total resistance = Sum of the individual resistance.

In such type of circuit all the lamps are controlled by the single switch and they cannot be controlled individually.The most common application of this circuit is for decoration purpose where a number of low voltage lamps are connected in series.

DC Parallel Circuit

The circuit which have DC source and one end of all the resistors is joined to a common point and other end are also joined to another common point so that current flows through them is called a DC parallel circuit.

The figure shows a simple parallel circuit. In this circuit the three resistor R_1, R_2, and R_3 are connected in parallel across a supply voltage of V volts. The current flowing through them is I_1, I_2 and I_3 respectively.

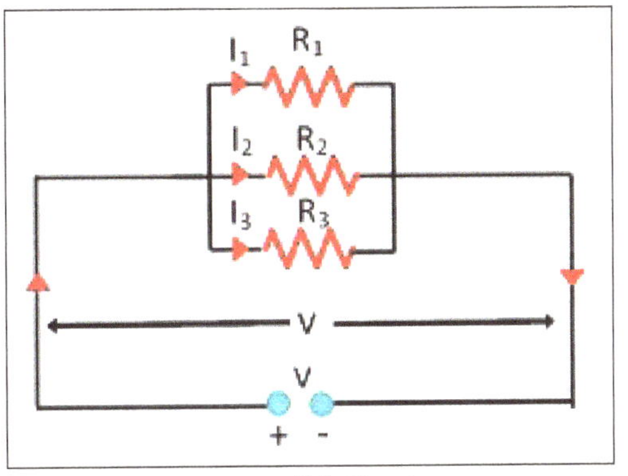

DC Parallel Circuit.

The total current drawn by the circuit,

$$I = I_1 + I_2 + I_3$$

$$I = \frac{V}{R_1} + \frac{V}{R_2} + \frac{V}{R_3}$$

Let R be the total or effective resistance of the circuit, then:

$$\frac{V}{R} = \frac{V}{R_1} + \frac{V}{R_2} + \frac{V}{R_3}$$

$$\frac{1}{R} = \frac{1}{R_1} + \frac{1}{R_2} + \frac{1}{R_3}$$

Reciprocal of total resistance = sum of reciprocal of the individual resistance.

All the resistance is operated to the same voltage, therefore all of them are connected in parallel. Each of them can be controlled individually with the help of a separate switch.

DC Series-parallel Circuit

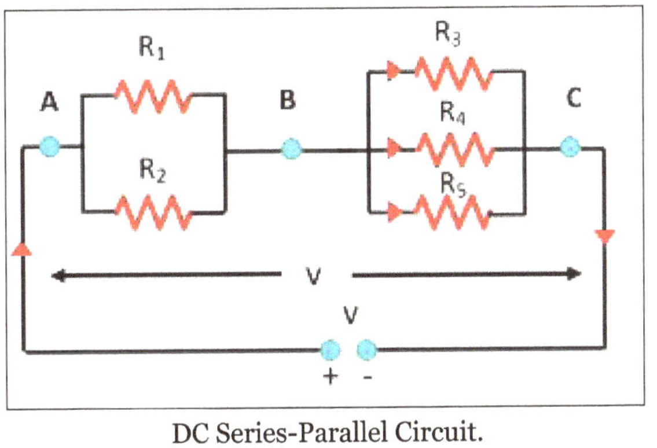

DC Series-Parallel Circuit.

The circuit in which series and parallel circuit are connected in series is called a series parallel circuit.The figure below show the series-parallel circuit. In this circuit, two resistor R_1 and R_2 are connected in parallel with each other across terminal AB. The other three resistors R_3, R_4 and R_6 are connected in parallel with each other across terminal BC.

The two groups of resistor R_{AB} and R_{BC} are connected in series with each other across the supply voltage of V volts. The total or effective resistance of the whole circuit can be determined as given below,

$$\frac{1}{R_{AB}} = \frac{1}{R_1} + \frac{1}{R_2} = \frac{R_1 + R_2}{R_1 R_2}$$

$$R_{AB} = \frac{R_1 R_2}{R_1 + R_2}$$

Similarly,

$$\frac{1}{R_{BC}} = \frac{1}{R_3} + \frac{1}{R_4} + \frac{1}{R_5} = \frac{R_3R_4 + R_4R_5 + R_5R_3}{R_3R_4R_5}$$

$$R_{BC} = \frac{R_3R_4R_5}{R_3R_4 + R_4R_5 + R_5R_3}$$

Total or effective resistance of the ciruit,

$$R = R_{AB} + R_{BC}$$

RL Circuit

A resistor–inductor circuit (RL circuit), or RL filter or RL network, is an electric circuit composed of resistors and inductors driven by a voltage or current source. A first-order RL circuit is composed of one resistor and one inductor and is the simplest type of RL circuit.

A first order RL circuit is one of the simplest analogue infinite impulse response electronic filters. It consists of a resistor and an inductor, either in series driven by a voltage source or in parallel driven by a current source.

The fundamental passive linear circuit elements are the resistor (R), capacitor (C) and inductor (L). These circuit elements can be combined to form an electrical circuit in four distinct ways: the RC circuit, the RL circuit, the LC circuit and the RLC circuit with the abbreviations indicating which components are used. These circuits exhibit important types of behaviour that are fundamental to analogue electronics. In particular, they are able to act as passive filters.

In practice, however, capacitors (and RC circuits) are usually preferred to inductors since they can be more easily manufactured and are generally physically smaller, particularly for higher values of components.

Both RC and RL circuits form a single-pole filter. Depending on whether the reactive element (C or L) is in series with the load, or parallel with the load will dictate whether the filter is low-pass or high-pass.

Frequently RL circuits are used for DC power supplies to RF amplifiers, where the inductor is used to pass DC bias current and block the RF getting back into the power supply.

Complex Impedance

The complex impedance Z_L (in ohms) of an inductor with inductance L (in henrys) is,

$$Z_L = Ls.$$

The complex frequency s is a complex number,

$$s = \sigma + j\omega,$$

where,

- j represents the imaginary unit: $j^2 = -1$,
- σ is the exponential decay constant (in radians per second), and
- ω is the angular frequency (in radians per second).

Eigenfunctions

The complex-valued eigenfunctions of *any* linear time-invariant (LTI) system are of the following forms:

$$\begin{aligned} \mathbf{V}(t) &= \mathbf{A}e^{st} = \mathbf{A}e^{(\sigma+j\omega)t} \\ \mathbf{A} &= Ae^{j\phi} \\ \Rightarrow \mathbf{V}(t) &= Ae^{j\phi}e^{(\sigma+j\omega)t} \\ &= Ae^{\sigma t}e^{j(\omega t+\phi)}. \end{aligned}$$

From Euler's formula, the real-part of these eigenfunctions are exponentially-decaying sinusoids:

$$v(t) = \mathrm{Re}\,V(t) = Ae^{\sigma t}\cos(\omega t + \phi).$$

Sinusoidal Steady State

Sinusoidal steady state is a special case in which the input voltage consists of a pure sinusoid (with no exponential decay). As a result,

$$\sigma = 0$$

and the evaluation of s becomes,

$$s = j\omega.$$

Series Circuit

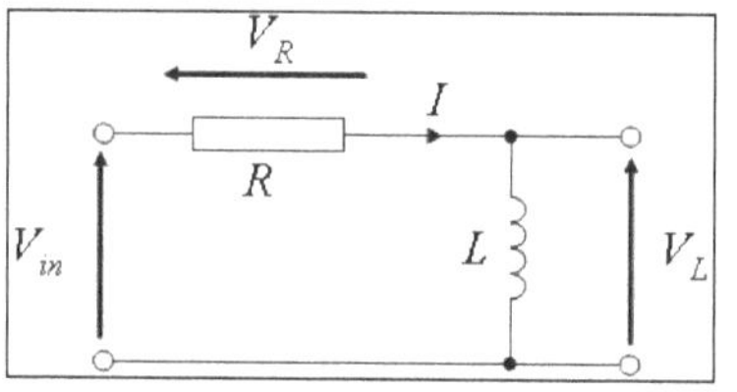

Series RL circuit.

By viewing the circuit as a voltage divider, we see that the voltage across the inductor is:

$$V_L(s) = \frac{Ls}{R + Ls} V_{\text{in}}(s),$$

and the voltage across the resistor is:

$$V_R(s) = \frac{R}{R + Ls} V_{\text{in}}(s).$$

Current

The current in the circuit is the same everywhere since the circuit is in series:

$$I(s) = \frac{V_{\text{in}}(s)}{R + Ls}.$$

Transfer Functions

The transfer function to the inductor voltage is

$$H_L(s) = \frac{V_L(s)}{V_{\text{in}}(s)} = \frac{Ls}{R + Ls} = G_L e^{j\phi_L}.$$

Similarly, the transfer function, to the resistor voltage is

$$H_R(s) = \frac{V_R(s)}{V_{\text{in}}(s)} = \frac{R}{R + Ls} = G_R e^{j\phi_R}.$$

The transfer function, to the current, is

$$H_I(s) = \frac{I(s)}{V_{\text{in}}(s)} = \frac{1}{R + Ls}.$$

Poles and Zeros

The transfer functions have a single pole located at

$$s = -\frac{R}{L}.$$

In addition, the transfer function for the inductor has a zero located at the origin.

Gain and Phase Angle

The gains across the two components are found by taking the magnitudes of the above expressions:

$$G_L = |H_L(\omega)| = \left|\frac{V_L(\omega)}{V_{\text{in}}(\omega)}\right| = \frac{\omega L}{\sqrt{R^2 + (\omega L)^2}}$$

and

$$G_R = |H_R(\omega)| = \left|\frac{V_R(\omega)}{V_{\text{in}}(\omega)}\right| = \frac{R}{\sqrt{R^2 + (\omega L)^2}},$$

and the phase angles are:

$$\phi_L = \angle H_L(s) = \tan^{-1}\left(\frac{R}{\omega L}\right)$$

and

$$\phi_R = \angle H_R(s) = \tan^{-1}\left(-\frac{\omega L}{R}\right).$$

Phasor Notation

These expressions together may be substituted into the usual expression for the phasor representing the output:

$$V_L = G_L V_{\text{in}} e^{j\phi_L}$$
$$V_R = G_R V_{\text{in}} e^{j\phi_R}$$

Impulse Response

The impulse response for each voltage is the inverse Laplace transform of the corresponding transfer function. It represents the response of the circuit to an input voltage consisting of an impulse or Dirac delta function.

The impulse response for the inductor voltage is

$$h_L(t) = \delta(t) - \frac{R}{L} e^{-t\frac{R}{L}} u(t) = \delta(t) - \frac{1}{\tau} e^{-\frac{t}{\tau}} u(t),$$

where $u(t)$ is the Heaviside step function and $\tau = L/R$ is the time constant.

Similarly, the impulse response for the resistor voltage is,

$$h_R(t) = \frac{R}{L} e^{-t\frac{R}{L}} u(t) = \frac{1}{\tau} e^{-\frac{t}{\tau}} u(t).$$

Zero-input Response

The zero-input response (ZIR), also called the natural response, of an RL circuit describes the behavior of the circuit after it has reached constant voltages and currents and is disconnected from any power source. It is called the zero-input response because it requires no input.

The ZIR of an RL circuit is:

$$I(t) = I(0) e^{-\frac{R}{L}t} = I(0) e^{-\frac{t}{\tau}}.$$

Frequency Domain Considerations

These are frequency domain expressions. Analysis of them will show which frequencies the circuits (or filters) pass and reject. This analysis rests on a consideration of what happens to these gains as the frequency becomes very large and very small.

As $\omega \to \infty$:

$$G_L \to 1 \quad \text{and} \quad G_R \to 0.$$

As $\omega \to 0$:

$$G_L \to 0 \quad \text{and} \quad G_R \to 1.$$

This shows that, if the output is taken across the inductor, high frequencies are passed and low frequencies are attenuated (rejected). Thus, the circuit behaves as a *high-pass filter*. If, though, the output is taken across the resistor, high frequencies are rejected and low frequencies are passed. In this configuration, the circuit behaves as a *low-pass filter*. Compare this with the behaviour of the resistor output in an RC circuit, where the reverse is the case.

The range of frequencies that the filter passes is called its bandwidth. The point at which the filter attenuates the signal to half its unfiltered power is termed its cutoff frequency. This requires that the gain of the circuit be reduced to,

$$G_L = G_R = \frac{1}{\sqrt{2}}.$$

Solving the above equation yields,

$$\omega_c = \frac{R}{L} \text{ rad/s} \quad \text{or} \quad f_c = \frac{R}{2\pi L} \text{ Hz},$$

which is the frequency that the filter will attenuate to half its original power.

Clearly, the phases also depend on frequency, although this effect is less interesting generally than the gain variations.

As $\omega \to 0$:

$$\phi_L \to 90^\circ = \frac{\pi}{2} \text{ radians} \quad \text{and} \quad \phi_R \to 0.$$

As $\omega \to \infty$:

$$\phi_L \to 0 \quad \text{and} \quad \phi_R \to -90^\circ = -\frac{\pi}{2} \text{ radians}.$$

So at DC (0 Hz), the resistor voltage is in phase with the signal voltage while the inductor voltage leads it by 90°. As frequency increases, the resistor voltage comes to have a 90° lag relative to the signal and the inductor voltage comes to be in-phase with the signal.

Time Domain Considerations

The most straightforward way to derive the time domain behaviour is to use the Laplace transforms of the expressions for V_L and V_R given above. This effectively transforms $j\omega \to s$. Assuming a step input (i.e., $V_{in} = 0$ before $t = 0$ and then $V_{in} = V$ afterwards):

$$V_{in}(s) = V \cdot \frac{1}{s}$$

$$V_L(s) = V \cdot \frac{sL}{R + sL} \cdot \frac{1}{s}$$

$$V_R(s) = V \cdot \frac{R}{R + sL} \cdot \frac{1}{s}.$$

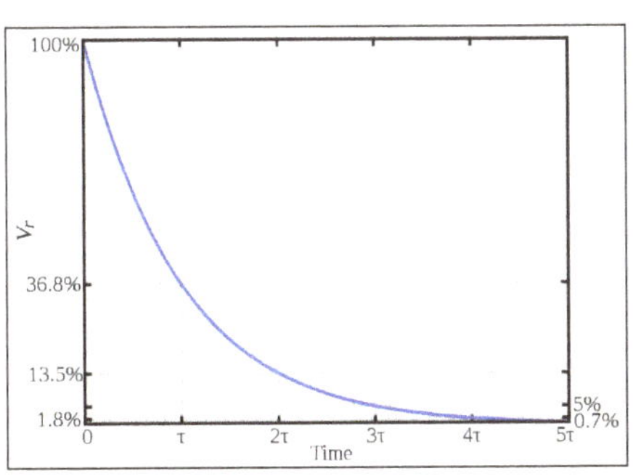

Inductor voltage step-response.

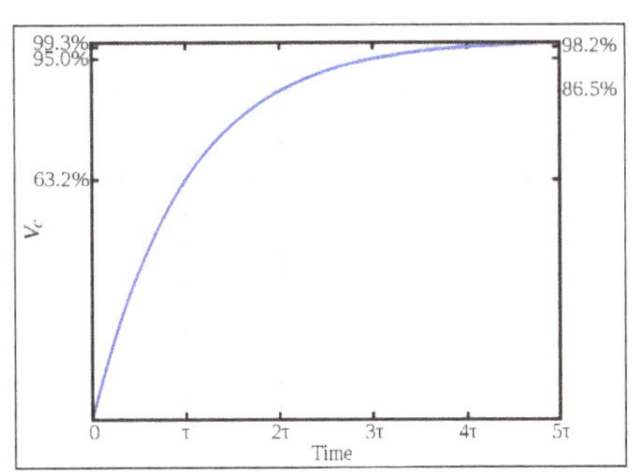

Resistor voltage step-response.

Partial fractions expansions and the inverse Laplace transform yield:

$$V_L(t) = Ve^{-t\frac{R}{L}}$$

$$V_R(t) = V\left(1 - e^{-t\frac{R}{L}}\right).$$

Thus, the voltage across the inductor tends towards 0 as time passes, while the voltage across the resistor tends towards V, as shown in the figures. This is in keeping with the intuitive point that the inductor will only have a voltage across as long as the current in the circuit is changing — as the circuit reaches its steady-state, there is no further current change and ultimately no inductor voltage.

These equations show that a series RL circuit has a time constant, usually denoted $\tau = L/R$ being the time it takes the voltage across the component to either fall (across the inductor) or rise (across the resistor) to within $1/e$ of its final value. That is, τ is the time it takes V_L to reach $V(1/e)$ and V_R to reach $V(1 - 1/e)$.

The rate of change is a *fractional* $1 - 1/e$ per τ. Thus, in going from $t = N\tau$ to $t = (N + 1)\tau$, the voltage will have moved about 63% of the way from its level at $t = N\tau$ toward its final value. So the voltage across the inductor will have dropped to about 37% after τ, and essentially to zero (0.7%) after about 5τ. Kirchhoff's voltage law implies that the voltage across the resistor will *rise* at the same rate. When the voltage source is then replaced with a short-circuit, the voltage across the resistor drops exponentially with t from V towards 0. The resistor will be discharged to about 37% after τ, and essentially fully discharged (0.7%) after about 5τ. Note that the current, I, in the circuit behaves as the voltage across the resistor does, via Ohm's Law.

The delay in the rise or fall time of the circuit is in this case caused by the back-EMF from the inductor which, as the current flowing through it tries to change, prevents the current (and hence the voltage across the resistor) from rising or falling much faster than the time-constant of the circuit. Since all wires have some self-inductance and resistance, all circuits have a time constant. As a result, when the power supply is switched on, the current does not instantaneously reach its steady-state value, V/R. The rise instead takes several time-constants to complete. If this were not the case, and the current were to reach steady-state immediately, extremely strong inductive electric fields would be generated by the sharp change in the magnetic field — this would lead to breakdown of the air in the circuit and electric arcing, probably damaging components (and users).

These results may also be derived by solving the differential equation describing the circuit:

$$V_{\text{in}} = IR + L\frac{dI}{dt}$$

$$V_R = V_{\text{in}} - V_L .$$

The first equation is solved by using an integrating factor and yields the current which must be differentiated to give V_L; the second equation is straightforward. The solutions are exactly the same as those obtained via Laplace transforms.

Short Circuit Equation

For short circuit evaluation, RL circuit is considered. The more general equation is:

$$v_{in}(t) = v_L(t) + v_R(t) = L\frac{di}{dt} + Ri$$

With initial condition:

$$i(0) = i_0$$

Which can be solved by Laplace transform:

$$V_{in}(s) = sLI - Li_0 + RI$$

Thus:

$$I(s) = \frac{Li_o + V_{in}}{sL + R}$$

Then antitransform returns:

$$i(t) = i_0 e^{-\frac{R}{L}t} + \mathcal{L}^{-1}\left[\frac{V_{in}}{sL + R}\right]$$

In case the source voltage is a Heaviside step function (DC):

$$v_{in}(t) = Eu(t)$$

Returns:

$$i(t) = i_0 e^{-\frac{R}{L}t} + \mathcal{L}^{-1}\left[\frac{E}{s(sL + R)}\right] = i_0 e^{-\frac{R}{L}t} + \frac{E}{R}\left(1 - e^{-\frac{R}{L}t}\right)$$

In case the source voltage is a sinusoidal function (AC):

$$v_{in}(t) = E\sin(\omega t) \Rightarrow V_{in}(s) = \frac{E\omega}{s^2 + \omega^2}$$

Returns:

$$i(t) = i_0 e^{-\frac{R}{L}t} + \mathcal{L}^{-1}\left[\frac{E\omega}{(s^2 + \omega^2)(sL + R)}\right] = i_0 e^{-\frac{R}{L}t} + \mathcal{L}^{-1}\left[\frac{E\omega}{2j\omega}\left(\frac{1}{s - j\omega} - \frac{1}{s + j\omega}\right)\frac{1}{(sL + R)}\right]$$

$$= i_0 e^{-\frac{R}{L}t} + \frac{E}{2jL}\mathcal{L}^{-1}\left[\frac{1}{s + \frac{R}{L}}\left(\frac{1}{\frac{R}{L} - j\omega} - \frac{1}{\frac{R}{L} + j\omega}\right) + \frac{1}{s - j\omega}\frac{1}{\frac{R}{L} + j\omega} - \frac{1}{s + j\omega}\frac{1}{\frac{R}{L} - j\omega}\right]$$

$$= i_0 e^{-\frac{R}{L}t} + \frac{E}{2jL}e^{-\frac{R}{L}t}2j\mathrm{Im}\left[\frac{1}{\frac{R}{L} - j\omega}\right] + \frac{E}{2jL}2j\mathrm{Im}\left[e^{j\omega t}\frac{1}{\frac{R}{L} + j\omega}\right]$$

$$= i_0 e^{-\frac{R}{L}t} + \frac{E\omega}{L\left(\left(\frac{R}{L}\right)^2 + \omega^2\right)} e^{-\frac{R}{L}t} + \frac{E}{L\left(\left(\frac{R}{L}\right)^2 + \omega^2\right)}\left(\frac{R}{L}\sin(\omega t) - \omega\cos(\omega t)\right)$$

$$i(t) = i_0 e^{-\frac{R}{L}t} + \frac{E\omega}{L\left(\left(\frac{R}{L}\right)^2 + \omega^2\right)} e^{-\frac{R}{L}t} + \frac{E}{L\sqrt{\left(\frac{R}{L}\right)^2 + \omega^2}}\sin\left(\omega t - \tan^{-1}\left(\frac{\omega L}{R}\right)\right)$$

Parallel Circuit

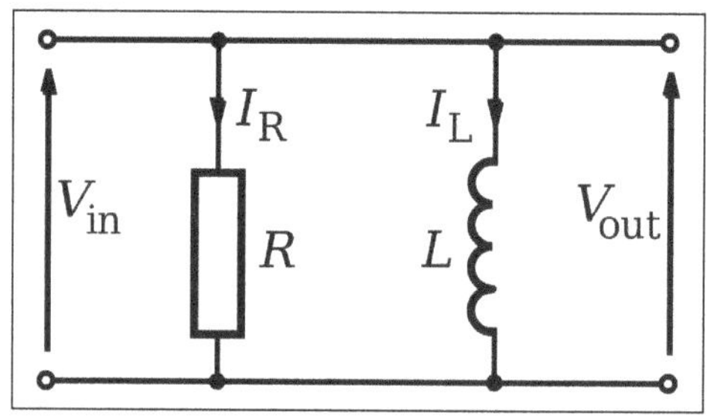

Parallel RL circuit.

The parallel RL circuit is generally of less interest than the series circuit unless fed by a current source. This is largely because the output voltage V_{out} is equal to the input voltage V_{in} — as a result, this circuit does not act as a filter for a voltage input signal.

With complex impedances:

$$I_R = \frac{V_{in}}{R}$$

$$I_L = \frac{V_{in}}{j\omega L} = -\frac{jV_{in}}{\omega L}.$$

This shows that the inductor lags the resistor (and source) current by 90°.

The parallel circuit is seen on the output of many amplifier circuits, and is used to isolate the amplifier from capacitive loading effects at high frequencies. Because of the phase shift introduced by capacitance, some amplifiers become unstable at very high frequencies, and tend to oscillate. This affects sound quality and component life (especially the transistors), and is to be avoided.

RC Circuit

A resistor–capacitor circuit (RC circuit), or RC filter or RC network, is an electric circuit composed of resistors and capacitors driven by a voltage or current source. A first order RC circuit is composed of one resistor and one capacitor and is the simplest type of RC circuit.

RC circuits can be used to filter a signal by blocking certain frequencies and passing others. The two most common RC filters are the high-pass filters and low-pass filters; band-pass filters and band-stop filters usually require RLC filters, though crude ones can be made with RC filters.

There are three basic, linear passive lumped analog circuit components: the resistor (R), the capacitor (C), and the inductor (L). These may be combined in the RC circuit, the RL circuit, the LC circuit, and the RLC circuit, with the acronyms indicating which components are used. These circuits, among them, exhibit a large number of important types of behaviour that are fundamental to much of analog electronics. In particular, they are able to act as passive filters.

Natural Response

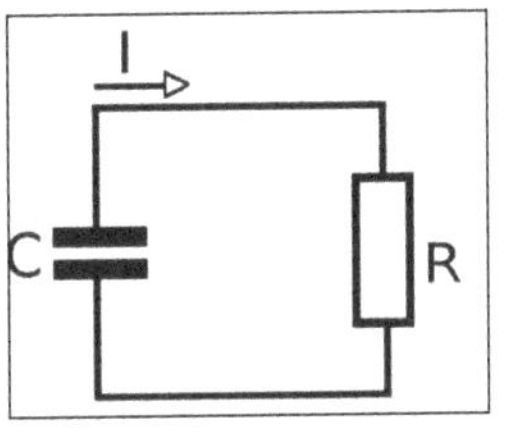

RC circuit.

The simplest RC circuit is a capacitor and a resistor in parallel. When a circuit consists of only a charged capacitor and a resistor, the capacitor will discharge its stored energy through the resistor. The voltage across the capacitor, which is time dependent, can be found by using Kirchhoff's current law, where the current discharging the capacitor must equal the current through the resistor. This results in the linear differential equation

$$C\frac{dV}{dt}+\frac{V}{R}=0,$$

where *C* is the capacitance of the capacitor.

Solving this equation for *V* yields the formula for exponential decay:

$$V(t)=V_0e^{-\frac{t}{RC}},$$

where V_0 is the capacitor voltage at time $t = 0$.

The time required for the voltage to fall to $\frac{V_0}{e}$ is called the RC time constant and is given by:

$$\tau=RC.$$

Where τ is measured in seconds, R in Ohms and C in Farads.

Complex Impedance

The complex impedance, Z_C (in ohms) of a capacitor with capacitance *C* (in farads) is

$$Z_C=\frac{1}{sC}$$

The complex frequency s is, in general, a complex number,

$$s = \sigma + j\omega,$$

where,

- j represents the imaginary unit: $j^2 = -1$,
- σ is the exponential decay constant (in nepers per second), and
- ω is the sinusoidal angular frequency (in radians per second).

Sinusoidal Steady State

Sinusoidal steady state is a special case in which the input voltage consists of a pure sinusoid (with no exponential decay). As a result, $\sigma = 0$, and the impedance becomes,

$$Z_C = -\frac{j}{\omega C}.$$

Series Circuit

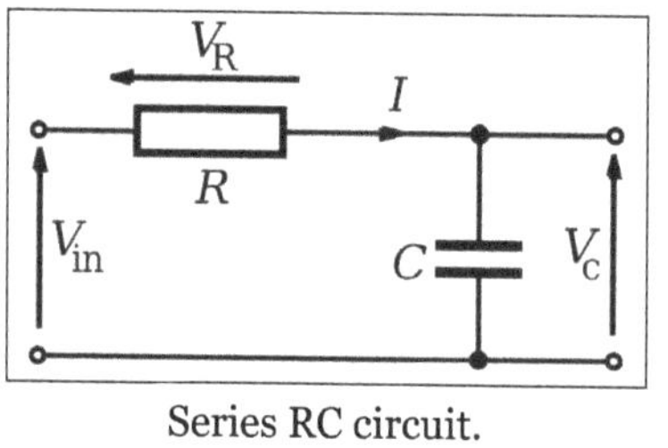

Series RC circuit.

By viewing the circuit as a voltage divider, the voltage across the capacitor is:

$$V_C(s) = \frac{\frac{1}{Cs}}{R + \frac{1}{Cs}} V_{\text{in}}(s) = \frac{1}{1 + RCs} V_{\text{in}}(s)$$

and the voltage across the resistor is:

$$V_R(s) = \frac{R}{R + \frac{1}{Cs}} V_{\text{in}}(s) = \frac{RCs}{1 + RCs} V_{\text{in}}(s).$$

Transfer Functions

The transfer function from the input voltage to the voltage across the capacitor is:

$$H_C(s) = \frac{V_C(s)}{V_{\text{in}}(s)} = \frac{1}{1 + RCs}.$$

Similarly, the transfer function from the input to the voltage across the resistor is:

$$H_R(s) = \frac{V_R(s)}{V_{\text{in}}(s)} = \frac{RCs}{1+RCs}.$$

Poles and Zeros

Both transfer functions have a single pole located at,

$$s = -\frac{1}{RC}.$$

In addition, the transfer function for the resistor has a zero located at the origin.

Gain and Phase

The magnitude of the gains across the two components are:

$$G_C = |H_C(j\omega)| = \left|\frac{V_C(j\omega)}{V_{\text{in}}(j\omega)}\right| = \frac{1}{\sqrt{1+(\omega RC)^2}}$$

and

$$G_R = |H_R(j\omega)| = \left|\frac{V_R(j\omega)}{V_{\text{in}}(j\omega)}\right| = \frac{\omega RC}{\sqrt{1+(\omega RC)^2}},$$

and the phase angles are,

$$\phi_C = \angle H_C(j\omega) = \tan^{-1}(-\omega RC)$$

and

$$\phi_R = \angle H_R(j\omega) = \tan^{-1}\left(\frac{1}{\omega RC}\right).$$

These expressions together may be substituted into the usual expression for the phasor representing the output:

$$V_C = G_C V_{\text{in}} e^{j\phi_C}$$
$$V_R = G_R V_{\text{in}} e^{j\phi_R}.$$

Current

The current in the circuit is the same everywhere since the circuit is in series:

$$I(s) = \frac{V_{\text{in}}(s)}{R+\frac{1}{Cs}} = \frac{Cs}{1+RCs} V_{\text{in}}(s).$$

Impulse Response

The impulse response for each voltage is the inverse Laplace transform of the corresponding transfer function. It represents the response of the circuit to an input voltage consisting of an impulse or Dirac delta function.

The impulse response for the capacitor voltage is,

$$h_C(t) = \frac{1}{RC} e^{-\frac{t}{RC}} u(t) = \frac{1}{\tau} e^{-\frac{t}{\tau}} u(t),$$

where $u(t)$ is the Heaviside step function and $\tau = RC$ is the time constant.

Similarly, the impulse response for the resistor voltage is,

$$h_R(t) = \delta(t) - \frac{1}{RC} e^{-\frac{t}{RC}} u(t) = \delta(t) - \frac{1}{\tau} e^{-\frac{t}{\tau}} u(t),$$

where $\delta(t)$ is the Dirac delta function.

Frequency-domain Considerations

These are frequency domain expressions. Analysis of them will show which frequencies the circuits (or filters) pass and reject. This analysis rests on a consideration of what happens to these gains as the frequency becomes very large and very small.

As $\omega \to \infty$:

$$G_C \to 0 \quad \text{and} \quad G_R \to 1.$$

As $\omega \to 0$:

$$G_C \to 1 \quad \text{and} \quad G_R \to 0.$$

This shows that, if the output is taken across the capacitor, high frequencies are attenuated (shorted to ground) and low frequencies are passed. Thus, the circuit behaves as a *low-pass filter*. If, though, the output is taken across the resistor, high frequencies are passed and low frequencies are attenuated (since the capacitor blocks the signal as its frequency approaches 0). In this configuration, the circuit behaves as a *high-pass filter*.

The range of frequencies that the filter passes is called its bandwidth. The point at which the filter attenuates the signal to half its unfiltered power is termed its cutoff frequency. This requires that the gain of the circuit be reduced to,

$$G_C = G_R = \frac{1}{\sqrt{2}}.$$

Solving the above equation yields,

$$\omega_c = \frac{1}{RC} \quad \text{or} \quad f_c = \frac{1}{2\pi RC}$$

which is the frequency that the filter will attenuate to half its original power.

Clearly, the phases also depend on frequency, although this effect is less interesting generally than the gain variations.

As $\omega \to 0$:

$$\phi_C \to 0 \quad \text{and} \quad \phi_R \to 90^\circ = \frac{\pi}{2} \text{ radians}.$$

As $\omega \to \infty$:

$$\phi_C \to -90^\circ = -\frac{\pi}{2} \text{ radians} \quad \text{and} \quad \phi_R \to 0.$$

So at DC (0 Hz), the capacitor voltage is in phase with the signal voltage while the resistor voltage leads it by 90°. As frequency increases, the capacitor voltage comes to have a 90° lag relative to the signal and the resistor voltage comes to be in-phase with the signal.

Time-domain Considerations

The most straightforward way to derive the time domain behaviour is to use the Laplace transforms of the expressions for V_C and V_R given above. This effectively transforms $j\omega \to s$. Assuming a step input (i.e. $V_{in} = 0$ before $t = 0$ and then $V_{in} = V$ afterwards):

$$V_{in}(s) = V \cdot \frac{1}{s}$$

$$V_C(s) = V \cdot \frac{1}{1+sRC} \cdot \frac{1}{s}$$

$$V_R(s) = V \cdot \frac{sRC}{1+sRC} \cdot \frac{1}{s}.$$

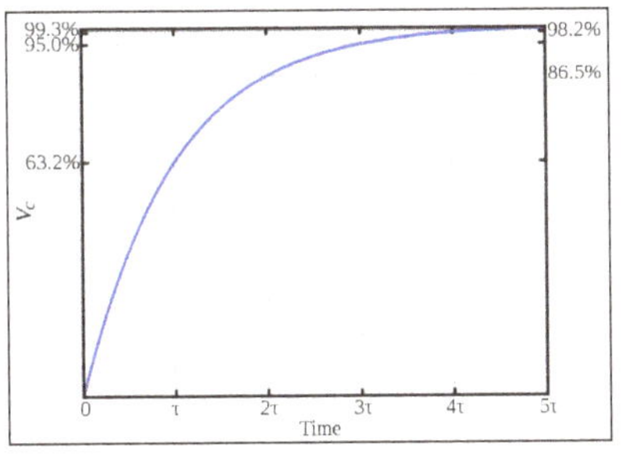

Capacitor voltage step-response.

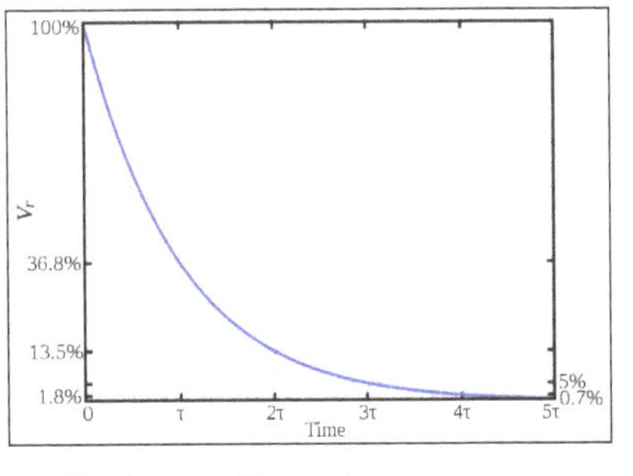

Resistor voltage step-response.

Partial fractions expansions and the inverse Laplace transform yield:

$$V_C(t) = V\left(1 - e^{-\frac{t}{RC}}\right)$$

$$V_R(t) = Ve^{-\frac{t}{RC}}.$$

These equations are for calculating the voltage across the capacitor and resistor respectively while the capacitor is charging; for discharging, the equations are vice versa. These equations can be rewritten in terms of charge and current using the relationships $C = Q/V$ and $V = IR$.

Thus, the voltage across the capacitor tends towards V as time passes, while the voltage across the resistor tends towards 0, as shown in the figures. This is in keeping with the intuitive point that the capacitor will be charging from the supply voltage as time passes, and will eventually be fully charged.

These equations show that a series RC circuit has a time constant, usually denoted $\tau = RC$ being the time it takes the voltage across the component to either rise (across the capacitor) or fall (across the resistor) to within $1/e$ of its final value. That is, τ is the time it takes V_C to reach $V(1 - 1/e)$ and V_R to reach $V(1/e)$.

The rate of change is a *fractional* $1 - 1/e$ per τ. Thus, in going from $t = N\tau$ to $t = (N + 1)\tau$, the voltage will have moved about 63.2% of the way from its level at $t = N\tau$ toward its final value. So the capacitor will be charged to about 63.2% after τ, and essentially fully charged (99.3%) after about 5τ. When the voltage source is replaced with a short-circuit, with the capacitor fully charged, the voltage across the capacitor drops exponentially with t from V towards 0. The capacitor will be discharged to about 36.8% after τ, and essentially fully discharged (0.7%) after about 5τ. Note that the current, I, in the circuit behaves as the voltage across the resistor does, via Ohm's Law.

These results may also be derived by solving the differential equations describing the circuit:

$$\frac{V_{\text{in}} - V_C}{R} = C\frac{dV_C}{dt}$$

$$V_R = V_{\text{in}} - V_C.$$

The first equation is solved by using an integrating factor and the second follows easily; the solutions are exactly the same as those obtained via Laplace transforms.

Integrator

Consider the output across the capacitor at *high* frequency, i.e.

$$\omega \gg \frac{1}{RC}.$$

This means that the capacitor has insufficient time to charge up and so its voltage is very small. Thus the input voltage approximately equals the voltage across the resistor. To see this, consider the expression for I given above:

$$I = \frac{V_{\text{in}}}{R + \frac{1}{j\omega C}},$$

but note that the frequency condition described means that,

$$\omega C \gg \frac{1}{R},$$

so,

$$I \approx \frac{V_{\text{in}}}{R}$$

which is just Ohm's Law.

Now,

$$V_C = \frac{1}{C}\int_0^t I\,dt,$$

so,

$$V_C \approx \frac{1}{RC}\int_0^t V_{in}\,dt,$$

which is an integrator *across the capacitor.*

Differentiator

Consider the output across the resistor at *low* frequency i.e.,

$$\omega \ll \frac{1}{RC}.$$

This means that the capacitor has time to charge up until its voltage is almost equal to the source's voltage. Considering the expression for I again, when,

$$R \ll \frac{1}{\omega C},$$

so

$$I \approx \frac{V_{\text{in}}}{\frac{1}{j\omega C}}$$

$$V_{\text{in}} \approx \frac{I}{j\omega C} = V_C.$$

Now,

$$V_R = IR = C\frac{dV_C}{dt}R$$

$$V_R \approx RC\frac{dV_{in}}{dt},$$

which is a differentiator *across the resistor.*

More accurate integration and differentiation can be achieved by placing resistors and capacitors as appropriate on the input and feedback loop of operational amplifiers.

Parallel Circuit

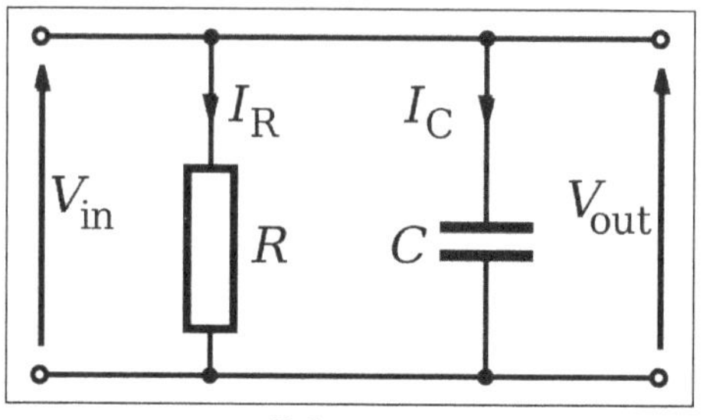

Parallel RC circuit.

The parallel RC circuit is generally of less interest than the series circuit. This is largely because the output voltage V_{out} is equal to the input voltage V_{in} — as a result, this circuit does not act as a filter on the input signal unless fed by a current source.

With complex impedances:

$$I_R = \frac{V_{in}}{R}$$
$$I_C = j\omega CV_{in}.$$

This shows that the capacitor current is 90° out of phase with the resistor (and source) current. Alternatively, the governing differential equations may be used:

$$I_R = \frac{V_{in}}{R}$$
$$I_C = C\frac{dV_{in}}{dt}.$$

When fed by a current source, the transfer function of a parallel RC circuit is:

$$\frac{V_{out}}{I_{in}} = \frac{R}{1+sRC}.$$

RLC Circuit

An RLC circuit is an electrical circuit consisting of a resistor (R), an inductor (L), and a capacitor (C), connected in series or in parallel. The name of the circuit is derived from the letters that are used to denote the constituent components of this circuit, where the sequence of the components may vary from RLC.

The circuit forms a harmonic oscillator for current, and resonates in a similar way as an LC circuit.

Introducing the resistor increases the decay of these oscillations, which is also known as damping. The resistor also reduces the peak resonant frequency. In ordinary conditions, some resistance is unavoidable even if a resistor is not specifically included as a component; an ideal, pure LC circuit exists only in the domain of superconductivity, a physical effect demonstrated to this point only at temperatures far below ambient temperatures found anywhere on the Earth's surface.

RLC circuits have many applications as oscillator circuits. Radio receivers and television sets use them for tuning to select a narrow frequency range from ambient radio waves. In this role, the circuit is often referred to as a tuned circuit. An RLC circuit can be used as a band-pass filter, band-stop filter, low-pass filter or high-pass filter. The tuning application, for instance, is an example of band-pass filtering. The RLC filter is described as a *second-order* circuit, meaning that any voltage or current in the circuit can be described by a second-order differential equation in circuit analysis.

The three circuit elements, R, L and C, can be combined in a number of different topologies. All three elements in series or all three elements in parallel are the simplest in concept and the most straightforward to analyse. There are, however, other arrangements, some with practical importance in real circuits. One issue often encountered is the need to take into account inductor resistance. Inductors are typically constructed from coils of wire, the resistance of which is not usually desirable, but it often has a significant effect on the circuit.

Basic Concepts

Resonance

An important property of this circuit is its ability to resonate at a specific frequency, the resonance frequency, f_0. Frequencies are measured in units of hertz. In this article, however, angular frequency, ω_0, is used which is more mathematically convenient. This is measured in radians per second. They are related to each other by a simple proportion,

$$\omega_0 = 2\pi f_0 .$$

Resonance occurs because energy is stored in two different ways: in an electric field as the capacitor is charged and in a magnetic field as current flows through the inductor. Energy can be transferred from one to the other within the circuit and this can be oscillatory. A mechanical analogy is a weight suspended on a spring which will oscillate up and down when released. This is no passing metaphor; a weight on a spring is described by exactly the same second order differential equation as an RLC circuit and for all the properties of the one system there will be found an analogous property of the other. The mechanical property answering to the resistor in the circuit is friction in the spring–weight system. Friction will slowly bring any oscillation to a halt if there is no external force driving it. Likewise, the resistance in an RLC circuit will "damp" the oscillation, diminishing it with time if there is no driving AC power source in the circuit.

The resonance frequency is defined as the frequency at which the impedance of the circuit is at a minimum. Equivalently, it can be defined as the frequency at which the impedance is purely real (that is, purely resistive). This occurs because the impedances of the inductor and capacitor at resonance are equal but of opposite sign and cancel out. Circuits where L and C are in parallel rather than series actually have a maximum impedance rather than a minimum impedance. For this reason they are often described as antiresonators, it is still usual, however, to name the frequency at which this occurs as the resonance frequency.

Natural Frequency

The resonance frequency is defined in terms of the impedance presented to a driving source. It is still possible for the circuit to carry on oscillating (for a time) after the driving source has been removed or it is subjected to a step in voltage (including a step down to zero). This is similar to the way that a tuning fork will carry on ringing after it has been struck, and the effect is often called ringing. This effect is the peak natural resonance frequency of the circuit and in general is not exactly the same as the driven resonance frequency, although the two will usually be quite close to each other. Various terms are used by different authors to distinguish the two, but resonance frequency unqualified usually means the driven resonance frequency. The driven frequency may be called the undamped resonance frequency or undamped natural frequency and the peak frequency may be called the damped resonance frequency or the damped natural frequency. The reason for this terminology is that the driven resonance frequency in a series or parallel resonant circuit has the value,

$$\omega_0 = \frac{1}{\sqrt{LC}}.$$

This is exactly the same as the resonance frequency of an LC circuit, that is, one with no resistor present. The resonant frequency for an RLC circuit is the same as a circuit in which there is no damping, hence undamped resonance frequency. The peak resonance frequency, on the other hand, depends on the value of the resistor and is described as the damped resonant frequency. A highly damped circuit will fail to resonate at all when not driven. A circuit with a value of resistor that causes it to be just on the edge of ringing is called critically damped. Either side of critically damped are described as underdamped (ringing happens) and overdamped (ringing is suppressed).

Circuits with topologies more complex than straightforward series or parallel have a driven resonance frequency that deviates from $\omega_0 = 1/\sqrt{LC}$, and for those the undamped resonance frequency, damped resonance frequency and driven resonance frequency can all be different.

Damping

Damping is caused by the resistance in the circuit. It determines whether or not the circuit will resonate naturally (that is, without a driving source). Circuits which will resonate in this way are described as underdamped and those that will not are overdamped. Damping attenuation (symbol α) is measured in nepers per second. However, the unitless damping factor (symbol ζ, zeta) is often a more useful measure, which is related to α by,

$$\zeta = \frac{\alpha}{\omega_0}.$$

The special case of $\zeta = 1$ is called critical damping and represents the case of a circuit that is just on the border of oscillation. It is the minimum damping that can be applied without causing oscillation.

Bandwidth

The resonance effect can be used for filtering, the rapid change in impedance near resonance can be used to pass or block signals close to the resonance frequency. A key parameter in filter design is bandwidth. The bandwidth is measured between the cutoff frequencies, most frequently defined as the frequencies at which the power passed through the circuit has fallen to half the value passed at resonance. There are two of these half-power frequencies, one above, and one below the resonance frequency,

$$\Delta\omega = \omega_2 - \omega_1.$$

where $\Delta\omega$ is the bandwidth, ω_1 is the lower half-power frequency and ω_2 is the upper half-power frequency. The bandwidth is related to attenuation by,

$$\Delta\omega = 2\alpha,$$

where the units are radians per second and nepers per second respectively. Other units may require a conversion factor. A more general measure of bandwidth is the fractional bandwidth, which expresses the bandwidth as a fraction of the resonance frequency and is given by,

The fractional bandwidth is also often stated as a percentage. The damping of filter circuits is adjusted to result in the required bandwidth. A narrow band filter, such as a notch filter, requires low damping. A wide band filter requires high damping.

Q factor

The Q factor is a widespread measure used to characterise resonators. It is defined as the peak energy stored in the circuit divided by the average energy dissipated in it per radian at resonance. Low-Q circuits are therefore damped and lossy and high-Q circuits are underdamped. Q is related to bandwidth; low-Q circuits are wide-band and high-Q circuits are narrow-band. In fact, it happens that Q is the inverse of fractional bandwidth,

$$Q = \frac{1}{B_\mathrm{f}} = \frac{\omega_0}{\Delta\omega}.$$

Q factor is directly proportional to selectivity, as the Q factor depends inversely on bandwidth.

For a series resonant circuit, the Q factor can be calculated as follows:

$$Q = \frac{1}{\omega_0 RC} = \frac{\omega_0 L}{R} = \frac{1}{R}\sqrt{\frac{L}{C}}.$$

Scaled Parameters

The parameters ζ, F_b, and Q are all scaled to ω_0. This means that circuits which have similar parameters share similar characteristics regardless of whether or not they are operating in the same frequency band.

Series Circuit

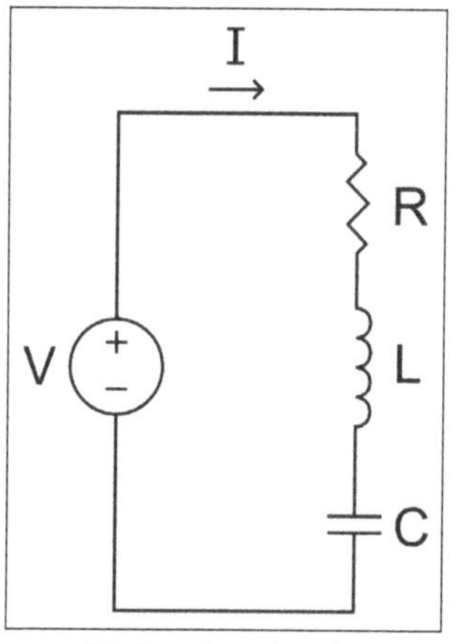

RLC series circuit.

- *V*, the voltage source powering the circuit.
- *I*, the current admitted through the circuit.
- *R*, the effective resistance of the combined load, source, and components.
- *L*, the inductance of the inductor component.
- *C*, the capacitance of the capacitor component.

In this circuit, the three components are all in series with the voltage source. The governing differential equation can be found by substituting into Kirchhoff's voltage law (KVL) the constitutive equation for each of the three elements. From the KVL,

$$V_R + V_L + V_C = V(t),$$

where V_R, V_L and V_C are the voltages across R, L and C respectively and $V(t)$ is the time-varying voltage from the source.

Substituting $V_R = RI(t)$, $V_L = L\frac{dI(t)}{dt}$ and $V_C = V(0) + \frac{1}{C}\int_0^t I(\tau)\,d\tau$ into the equation above yields:

$$RI(t) + L\frac{dI(t)}{dt} + V(0) + \frac{1}{C}\int_0^t I(\tau)\,d\tau = V(t).$$

For the case where the source is an unchanging voltage, taking the time derivative and dividing by *L* leads to the following second order differential equation:

$$\frac{d^2}{dt^2}I(t) + \frac{R}{L}\frac{d}{dt}I(t) + \frac{1}{LC}I(t) = 0.$$

This can usefully be expressed in a more generally applicable form:

$$\frac{d^2}{dt^2}I(t)+2\alpha\frac{d}{dt}I(t)+\omega_0^2 I(t)=0.$$

α and ω_0 are both in units of angular frequency. α is called the *neper frequency*, or *attenuation*, and is a measure of how fast the transient response of the circuit will die away after the stimulus has been removed. Neper occurs in the name because the units can also be considered to be nepers per second, neper being a unit of attenuation. ω_0 is the angular resonance frequency.

For the case of the series RLC circuit these two parameters are given by:

$$\alpha=\frac{R}{2L}$$

$$\omega_0=\frac{1}{\sqrt{LC}}.$$

A useful parameter is the *damping factor*, ζ, which is defined as the ratio of these two; although, sometimes α is referred to as the damping factor and ζ is not used.

$$\zeta=\frac{\alpha}{\omega_0}.$$

In the case of the series RLC circuit, the damping factor is given by,

$$\zeta=\frac{R}{2}\sqrt{\frac{C}{L}}.$$

The value of the damping factor determines the type of transient that the circuit will exhibit.

Transient Response

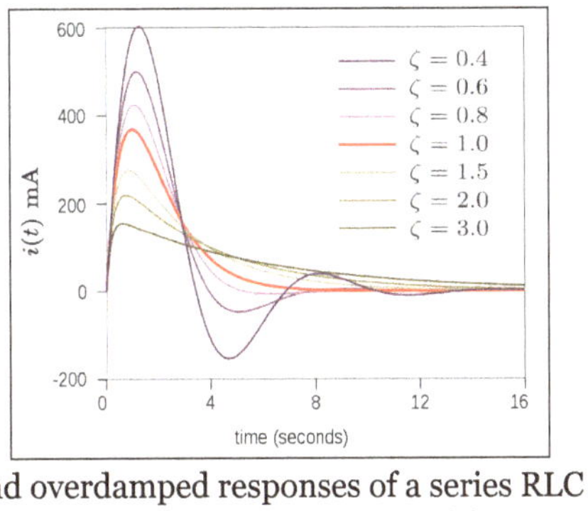

Plot showing underdamped and overdamped responses of a series RLC circuit. The critical damping plot is the bold red curve. The plots are normalised for $L = 1$, $C = 1$ and $\omega_0 = 1$.

The differential equation for the circuit solves in three different ways depending on the value of ζ. These are underdamped ($\zeta < 1$), overdamped ($\zeta > 1$) and critically damped ($\zeta = 1$). The differential equation has the characteristic equation,

$$s^2+2\alpha s+\omega_0^2=0.$$

The roots of the equation in *s* are,

$$s_1 = -\alpha + \sqrt{\alpha^2 - \omega_0^2}$$
$$s_2 = -\alpha - \sqrt{\alpha^2 - \omega_0^2}.$$

The general solution of the differential equation is an exponential in either root or a linear superposition of both,

$$I(t) = A_1 e^{s_1 t} + A_2 e^{s_2 t}.$$

The coefficients A_1 and A_2 are determined by the boundary conditions of the specific problem being analysed. That is, they are set by the values of the currents and voltages in the circuit at the onset of the transient and the presumed value they will settle to after infinite time.

Overdamped Response

The overdamped response ($\zeta > 1$) is,

$$I(t) = A_1 e^{-\omega_0\left(\zeta + \sqrt{\zeta^2 - 1}\right)t} + A_2 e^{-\omega_0\left(\zeta - \sqrt{\zeta^2 - 1}\right)t}.$$

The overdamped response is a decay of the transient current without oscillation.

Underdamped Response

The underdamped response ($\zeta < 1$) is,

$$I(t) = B_1 e^{-\alpha t}\cos(\omega_d t) + B_2 e^{-\alpha t}\sin(\omega_d t).$$

By applying standard trigonometric identities the two trigonometric functions may be expressed as a single sinusoid with phase shift,

$$I(t) = B_3 e^{-\alpha t}\sin(\omega_d t + \varphi).$$

The underdamped response is a decaying oscillation at frequency ω_d. The oscillation decays at a rate determined by the attenuation α. The exponential in α describes the envelope of the oscillation. B_1 and B_2 (or B_3 and the phase shift φ in the second form) are arbitrary constants determined by boundary conditions. The frequency ω_d is given by,

$$\omega_d = \sqrt{\omega_0^2 - \alpha^2} = \omega_0\sqrt{1 - \zeta^2}.$$

This is called the damped resonance frequency or the damped natural frequency. It is the frequency the circuit will naturally oscillate at if not driven by an external source. The resonance frequency, ω_0, which is the frequency at which the circuit will resonate when driven by an external oscillation, may often be referred to as the undamped resonance frequency to distinguish it.

Critically Damped Response

The critically damped response ($\zeta = 1$) is,

$$I(t) = D_1 te^{-\alpha t} + D_2 e^{-\alpha t}.$$

The critically damped response represents the circuit response that decays in the fastest possible time without going into oscillation. This consideration is important in control systems where it is required to reach the desired state as quickly as possible without overshooting. D_1 and D_2 are arbitrary constants determined by boundary conditions.

Laplace Domain

The series RLC can be analyzed for both transient and steady AC state behavior using the Laplace transform. If the voltage source above produces a waveform with Laplace-transformed *V*(*s*) (where *s* is the complex frequency $s = \sigma + j\omega$), the KVL can be applied in the Laplace domain:

$$V(s) = I(s)\left(R + Ls + \frac{1}{Cs}\right),$$

where *I*(*s*) is the Laplace-transformed current through all components. Solving for *I*(*s*):

$$I(s) = \frac{1}{R + Ls + \frac{1}{Cs}} V(s).$$

And rearranging, we have,

$$I(s) = \frac{s}{L\left(s^2 + \frac{R}{L}s + \frac{1}{LC}\right)} V(s).$$

Laplace Admittance

Solving for the Laplace admittance *Y*(*s*):

$$Y(s) = \frac{I(s)}{V(s)} = \frac{s}{L\left(s^2 + \frac{R}{L}s + \frac{1}{LC}\right)}.$$

Simplifying using parameters α and ω_0, we have,

$$Y(s) = \frac{I(s)}{V(s)} = \frac{s}{L\left(s^2 + 2\alpha s + \omega_0^2\right)}.$$

Poles and Zeros

The zeros of $Y(s)$ are those values of s such that $Y(s) = 0$:

$$s = 0 \quad \text{and} \quad |s| \to \infty.$$

The poles of $Y(s)$ are those values of s such that $Y(s) \to \infty$. By the quadratic formula, we find,

$$s = -\alpha \pm \sqrt{\alpha^2 - \omega_0^2}.$$

The poles of $Y(s)$ are identical to the roots s_1 and s_2 of the characteristic polynomial of the differential equation.

General Solution

For an arbitrary $V(t)$, the solution obtained by inverse transform of $I(s)$ is:

- In the underdamped case, $\omega_0 > \alpha$:

$$I(t) = \frac{1}{L}\int_0^t V(t-\tau)e^{-\alpha\tau}\left(\cos\omega_d\tau - \frac{\alpha}{\omega_d}\sin\omega_d\tau\right)d\tau,$$

- In the critically damped case, $\omega_0 = \alpha$:

$$I(t) = \frac{1}{L}\int_0^t V(t-\tau)e^{-\alpha\tau}(1-\alpha\tau)\,d\tau,$$

- In the overdamped case, $\omega_0 < \alpha$:

$$I(t) = \frac{1}{L}\int_0^t V(t-\tau)e^{-\alpha\tau}\left(\cosh\omega_r\tau - \frac{\alpha}{\omega_r}\sinh\omega_r\tau\right)d\tau,$$

where $\omega_r = \sqrt{\alpha^2 - \omega_0^2}$, and cosh and sinh are the usual hyperbolic functions.

Sinusoidal Steady State

Sinusoidal steady state is represented by letting $s = j\omega$, where j is the imaginary unit. Taking the magnitude of the above equation with this substitution:

$$|Y(j\omega)| = \frac{1}{\sqrt{R^2 + \left(\omega L - \dfrac{1}{\omega C}\right)^2}}.$$

and the current as a function of ω can be found from

$$|I(j\omega)| = |Y(j\omega)| \cdot |V(j\omega)|.$$

There is a peak value of $|I(j\omega)|$. The value of ω at this peak is, in this particular case, equal to the undamped natural resonance frequency:

$$\omega_0 = \frac{1}{\sqrt{LC}}.$$

Parallel Circuit

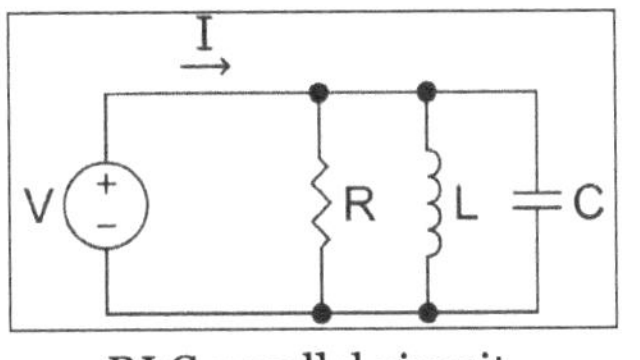

RLC parallel circuit.

- V – the voltage source powering the circuit.
- I – the current admitted through the circuit.
- R – the equivalent resistance of the combined source, load, and components.
- L – the inductance of the inductor component.
- C – the capacitance of the capacitor component.

The properties of the parallel RLC circuit can be obtained from the duality relationship of electrical circuits and considering that the parallel RLC is the dual impedance of a series RLC. Considering this, it becomes clear that the differential equations describing this circuit are identical to the general form of those describing a series RLC.

For the parallel circuit, the attenuation α is given by,

$$\alpha = \frac{1}{2RC}$$

and the damping factor is consequently,

$$\zeta = \frac{1}{2R}\sqrt{\frac{L}{C}}.$$

Likewise, the other scaled parameters, fractional bandwidth and Q are also reciprocals of each other. This means that a wide-band, low-Q circuit in one topology will become a narrow-band, high-Q circuit in the other topology when constructed from components with identical values. The fractional bandwidth and Q of the parallel circuit are given by,

$$B_{\mathrm{f}} = \frac{1}{R}\sqrt{\frac{L}{C}}$$

$$Q = R\sqrt{\frac{C}{L}}.$$

Notice that the formulas here are the reciprocals of the formulas for the series circuit.

Frequency Domain

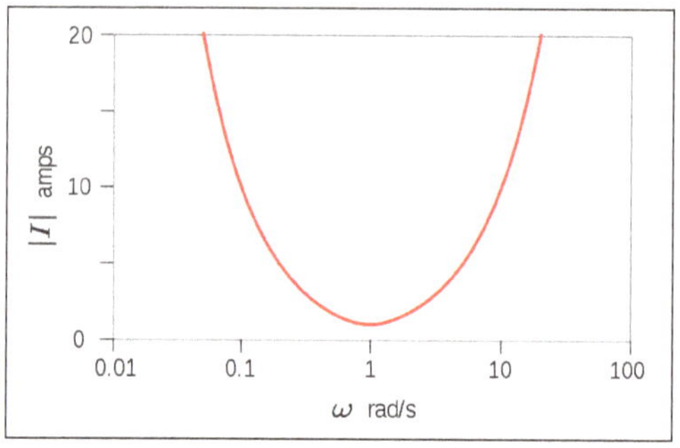

Sinusoidal steady-state analysis. Normalised to $R = 1\ \Omega$, $C = 1$ F, $L = 1$ H, and $V = 1$ V.

The complex admittance of this circuit is given by adding up the admittances of the components:

$$\frac{1}{Z} = \frac{1}{Z_L} + \frac{1}{Z_C} + \frac{1}{Z_R}$$
$$= \frac{1}{j\omega L} + j\omega C + \frac{1}{R}.$$

The change from a series arrangement to a parallel arrangement results in the circuit having a peak in impedance at resonance rather than a minimum, so the circuit is an anti-resonator.

The graph opposite shows that there is a minimum in the frequency response of the current at the resonance frequency $\omega_0 = 1/\sqrt{LC}$ when the circuit is driven by a constant voltage. On the other hand, if driven by a constant current, there would be a maximum in the voltage which would follow the same curve as the current in the series circuit.

Other Configurations

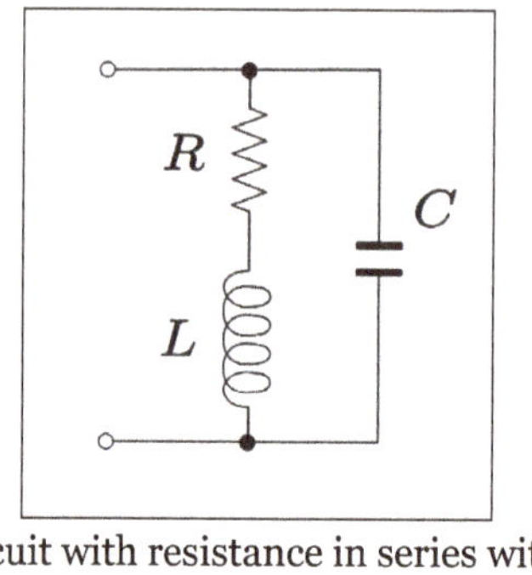

RLC parallel circuit with resistance in series with the inductor.

A series resistor with the inductor in a parallel LC circuit as shown in figure is a topology commonly encountered where there is a need to take into account the resistance of the coil winding. Parallel LC circuits are frequently used for bandpass filtering and the Q is largely governed by this resistance. The resonant frequency of this circuit is,

$$\omega_0 = \sqrt{\frac{1}{LC} - \left(\frac{R}{L}\right)^2}.$$

This is the resonant frequency of the circuit defined as the frequency at which the admittance has zero imaginary part. The frequency that appears in the generalised form of the characteristic equation,

$$s^2 + 2\alpha s + \omega_0'^2 = 0$$

is not the same frequency. In this case it is the natural undamped resonant frequency:

$$\omega_0' = \sqrt{\frac{1}{LC}}.$$

The frequency ω_m at which the impedance magnitude is maximum is given by,

$$\omega_m = \omega_0' \sqrt{-\frac{1}{Q_L^2} + \sqrt{1 + \frac{2}{Q_L^2}}},$$

where $Q_L = \omega'_0 L/R$ is the quality factor of the coil. This can be well approximated by,

$$\omega_m \approx \omega_0' \sqrt{1 - \frac{1}{2Q_L^4}}.$$

Furthermore, the exact maximum impedance magnitude is given by,

$$|Z|_{max} = RQ_L^2 \sqrt{\frac{1}{2Q_L\sqrt{Q_L^2+2} - 2Q_L^2 - 1}}.$$

For values of Q_L greater than unity, this can be well approximated by,

$$|Z|_{max} \approx RQ_L^2.$$

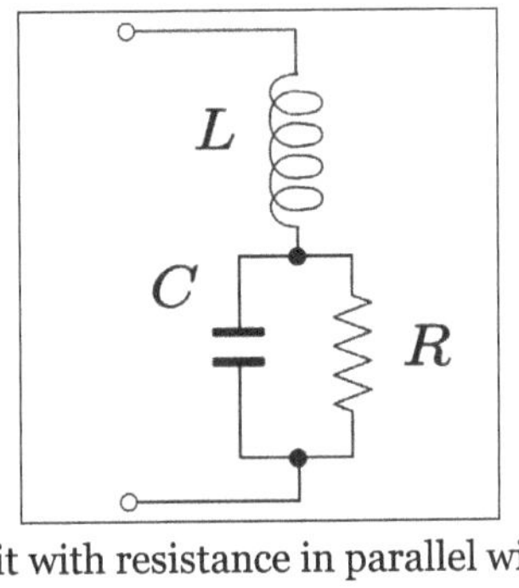

RLC series circuit with resistance in parallel with the capacitor.

In the same vein, a resistor in parallel with the capacitor in a series LC circuit can be used to represent a capacitor with a lossy dielectric. This configuration is shown in figure. The resonant frequency (frequency at which the impedance has zero imaginary part) in this case is given by,

$$\omega_0 = \sqrt{\frac{1}{LC} - \frac{1}{(RC)^2}},$$

while the frequency ω_m at which the impedance magnitude is maximum is given by,

$$\omega_m = \omega_0' \sqrt{-\frac{1}{Q_C^2} + \sqrt{1 + \frac{2}{Q_C^2}}},$$

where, $Q_C = \omega'_0 RC$.

Applications

Variable Tuned Circuits

A very frequent use of these circuits is in the tuning circuits of analogue radios. Adjustable tuning is commonly achieved with a parallel plate variable capacitor which allows the value of C to be changed and tune to stations on different frequencies. For the IF stage in the radio where the tuning is preset in the factory, the more usual solution is an adjustable core in the inductor to adjust L. In this design, the core (made of a high permeability material that has the effect of increasing inductance) is threaded so that it can be screwed further in, or screwed further out of the inductor winding as required.

Filters

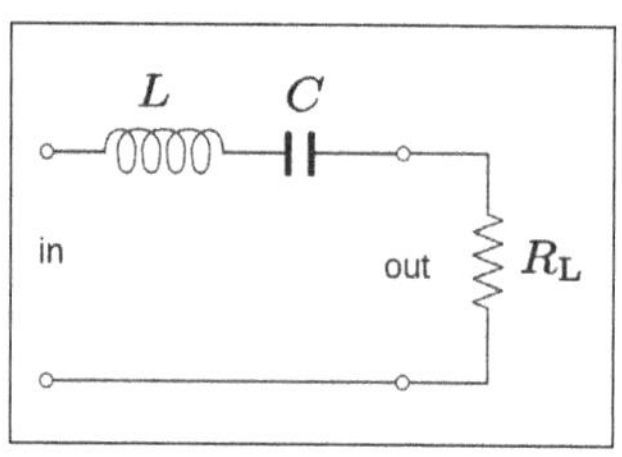

RLC circuit as a low-pass filter.

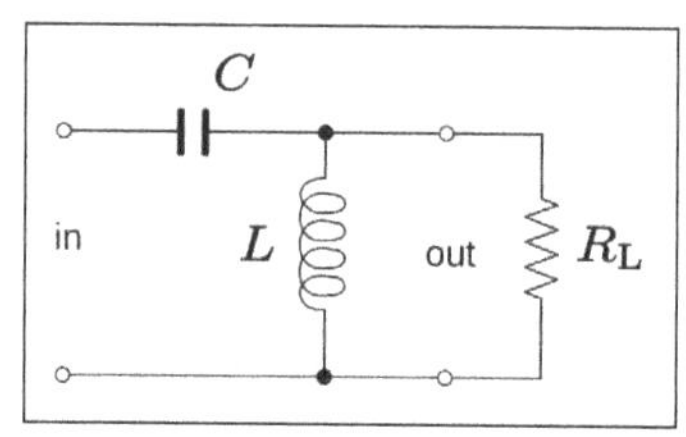

RLC circuit as a high-pass filter.

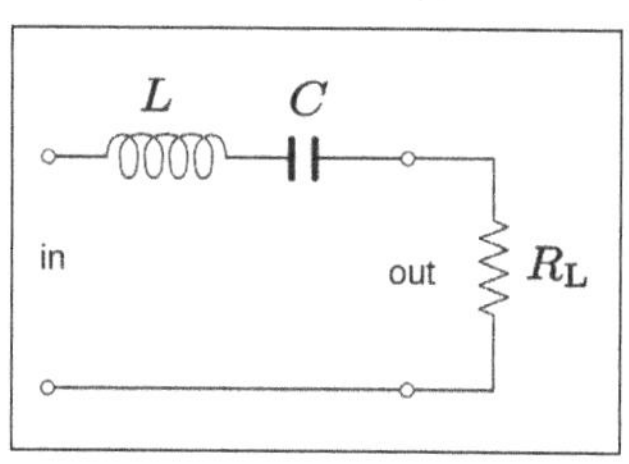

RLC circuit as a series band-pass filter in series with the line.

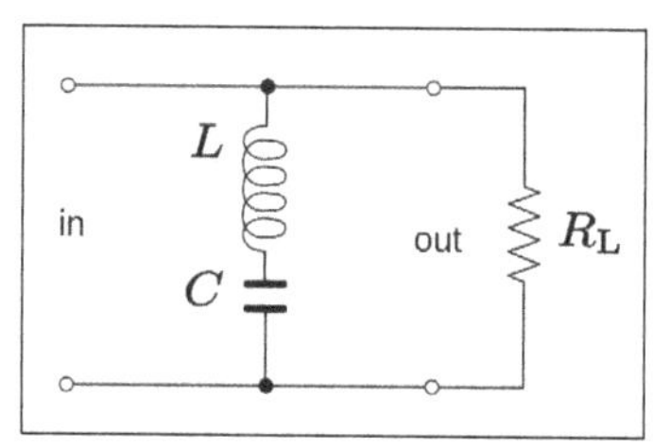

RLC circuit as a parallel band-pass filter in shunt across the line.

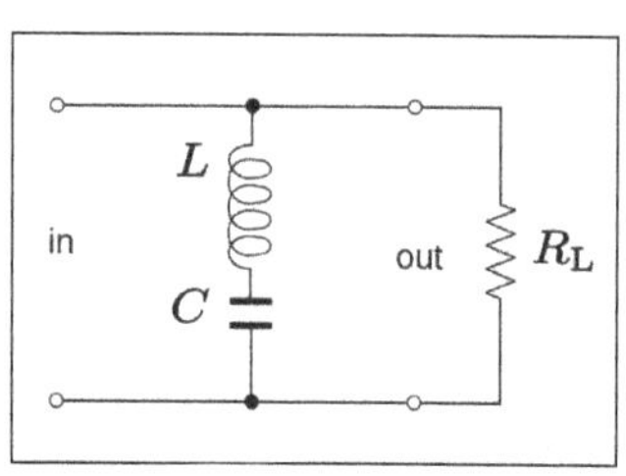

RLC circuit as a series band-stop filter in shunt across the line.

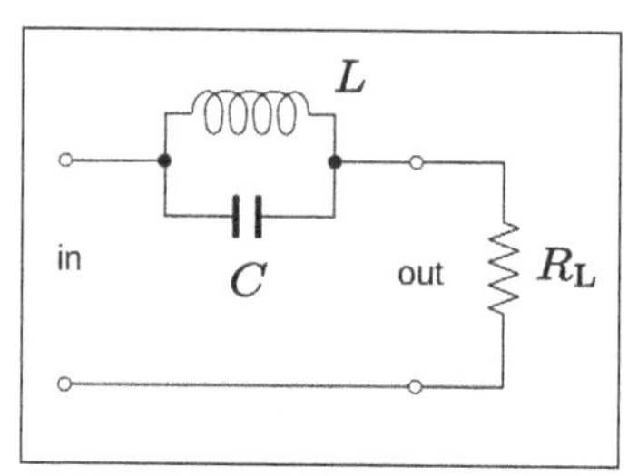

RLC circuit as a parallel band-stop filter in series with the line.

In the filtering application, the resistor becomes the load that the filter is working into. The value of the damping factor is chosen based on the desired bandwidth of the filter. For a wider bandwidth, a larger value of the damping factor is required (and vice versa). The three components give the designer three degrees of freedom. Two of these are required to set the bandwidth and resonant frequency. The designer is still left with one which can be used to scale *R*, *L* and *C* to convenient practical values. Alternatively, *R* may be predetermined by the external circuitry which will use the last degree of freedom.

Low-pass Filter

An RLC circuit can be used as a low-pass filter. The circuit configuration is shown in figure. The corner frequency, that is, the frequency of the 3 dB point, is given by,

$$\omega_c = \frac{1}{\sqrt{LC}}.$$

This is also the bandwidth of the filter. The damping factor is given by,

$$\zeta = \frac{1}{2R_L}\sqrt{\frac{L}{C}}.$$

High-pass Filter

A high-pass filter is shown in figure. The corner frequency is the same as the low-pass filter:

$$\omega_c = \frac{1}{\sqrt{LC}}.$$

The filter has a stop-band of this width.

Band-pass Filter

A band-pass filter can be formed with an RLC circuit by either placing a series LC circuit in series with the load resistor or else by placing a parallel LC circuit in parallel with the load resistor. These arrangements are shown in figures and respectively. The centre frequency is given by,

$$\omega_c = \frac{1}{\sqrt{LC}},$$

and the bandwidth for the series circuit is,

$$\Delta\omega = \frac{R_L}{L}.$$

The shunt version of the circuit is intended to be driven by a high impedance source, that is, a constant current source. Under those conditions the bandwidth is,

$$\Delta\omega = \frac{1}{CR_L}.$$

Band-stop Filter

Figure shows a band-stop filter formed by a series LC circuit in shunt across the load. Figure is a band-stop filter formed by a parallel LC circuit in series with the load. The first case requires a high impedance source so that the current is diverted into the resonator when it becomes low impedance at resonance. The second case requires a low impedance source so that the voltage is dropped across the antiresonator when it becomes high impedance at resonance.

Oscillators

For applications in oscillator circuits, it is generally desirable to make the attenuation (or equivalently, the damping factor) as small as possible. In practice, this objective requires making the circuit's resistance R as small as physically possible for a series circuit, or alternatively increasing R to as much as possible for a parallel circuit. In either case, the RLC circuit becomes a good approximation to an ideal LC circuit. However, for very low-attenuation circuits (high Q-factor), issues such as dielectric losses of coils and capacitors can become important.

In an oscillator circuit,

$$\alpha \ll \omega_0 ,$$

or equivalently,

$$\zeta \ll 1.$$

As a result,

$$\omega_\mathrm{d} \approx \omega_0 .$$

Voltage Multiplier

In a series RLC circuit at resonance, the current is limited only by the resistance of the circuit,

$$I = \frac{V}{R}.$$

If R is small, consisting only of the inductor winding resistance say, then this current will be large. It will drop a voltage across the inductor of,

$$V_L = \frac{V}{R}\omega_0 L.$$

An equal magnitude voltage will also be seen across the capacitor but in antiphase to the inductor. If R can be made sufficiently small, these voltages can be several times the input voltage. The voltage ratio is, in fact, the Q of the circuit,

$$\frac{V_L}{V} = Q.$$

A similar effect is observed with currents in the parallel circuit. Even though the circuit appears as high impedance to the external source, there is a large current circulating in the internal loop of the parallel inductor and capacitor.

Pulse Discharge Circuit

An overdamped series RLC circuit can be used as a pulse discharge circuit. Often it is useful to know the values of components that could be used to produce a waveform. This is described by the form,

$$I(t) = I_0\left(e^{-\alpha t} - e^{-\beta t}\right).$$

Such a circuit could consist of an energy storage capacitor, a load in the form of a resistance, some circuit inductance and a switch – all in series. The initial conditions are that the capacitor is at voltage, V_0, and there is no current flowing in the inductor. If the inductance L is known, then the remaining parameters are given by the following – capacitance:

$$C = \frac{1}{L\alpha\beta},$$

resistance (total of circuit and load):

$$R = L(\alpha + \beta),$$

initial terminal voltage of capacitor:

$$V_0 = -I_0 L\alpha\beta\left(\frac{1}{\beta} - \frac{1}{\alpha}\right).$$

Rearranging for the case where R is known – capacitance:

$$C = \frac{\alpha + \beta}{R\alpha\beta},$$

inductance (total of circuit and load):

$$L = \frac{R}{\alpha + \beta},$$

initial terminal voltage of capacitor:

$$V_0 = \frac{-I_0 R\alpha\beta}{\alpha + \beta}\left(\frac{1}{\beta} - \frac{1}{\alpha}\right).$$

Wiring Diagram

A wiring diagram is a simplified conventional pictorial representation of an electrical circuit. It shows the components of the circuit as simplified shapes, and the power and signal connections between the devices.

A wiring diagram usually gives information about the relative position and arrangement of devices and terminals on the devices, to help in building or servicing the device. This is unlike a schematic diagram, where the arrangement of the components' interconnections on the diagram usually does not correspond to the components' physical locations in the finished device. A pictorial diagram would show more detail of the physical appearance, whereas a wiring diagram uses a more symbolic notation to emphasize interconnections over physical appearance. A wiring diagram is often used to troubleshoot problems and to make sure that all the connections have been made and that everything is present.

Architectural Wiring Diagrams

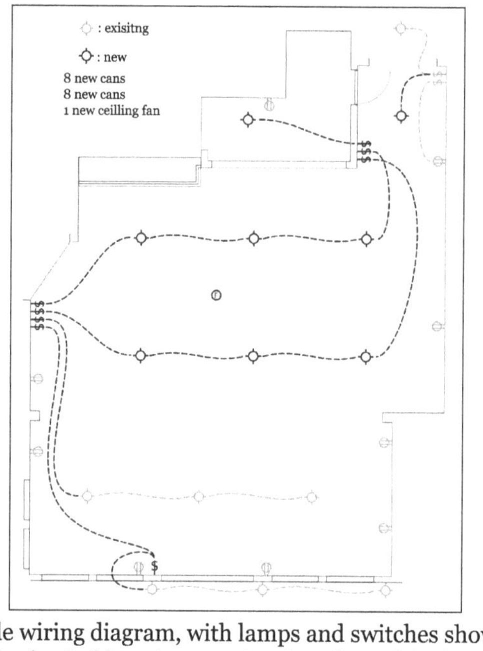

Architectural style wiring diagram, with lamps and switches shown symbolically in their physical locations on the plan view of the building.

Architectural wiring diagrams show the approximate locations and interconnections of receptacles, lighting, and permanent electrical services in a building. Interconnecting wire routes may be shown approximately, where particular receptacles or fixtures must be on a common circuit.

Wiring diagrams use standard symbols for wiring devices, usually different from those used on schematic diagrams. The electrical symbols not only show where something is to be installed, but also what type of device is being installed. For example, a surface ceiling light is shown by one symbol, a recessed ceiling light has a different symbol, and a surface fluorescent light has another symbol. Each type of switch has a different symbol and so do the various outlets. There are symbols that show the location of smoke detectors, the doorbell chime, and thermostat. On large projects symbols may be numbered to show, for example, the panel board and circuit to which the device connects, and also to identify which of several types of fixture are to be installed at that location.

A set of wiring diagrams may be required by the electrical inspection authority to approve connection of the residence to the public electrical supply system. Wiring diagrams will also include panel schedules for circuit breaker panelboards, and riser diagrams for special services such as fire alarm or closed circuit television or other special services.

Electrical Circuit Breaker

A circuit breaker is an automatically operated electrical switch designed to protect an electrical circuit from damage caused by excess current from an overload or short circuit. Its basic function is to interrupt current flow after a fault is detected. Unlike a fuse, which operates once and then must be replaced, a circuit breaker can be reset (either manually or automatically) to resume normal operation.

Circuit breakers are made in varying sizes, from small devices that protect low-current circuits or individual household appliance, up to large switchgear designed to protect high voltage circuits feeding an entire city. The generic function of a circuit breaker, RCD or a fuse, as an automatic means of removing power from a faulty system is often abbreviated as OCPD (Over Current Protection Device).

An early form of circuit breaker was described by Thomas Edison in an 1879 patent application, although his commercial power distribution system used fuses. Its purpose was to protect lighting circuit wiring from accidental short circuits and overloads. A modern miniature circuit breaker similar to the ones now in use was patented by Brown, Boveri & Cie in 1924. Hugo Stotz, an engineer who had sold his company to BBC, was credited as the inventor on DRP (*Deutsches Reichspatent*) 458392. Stotz's invention was the forerunner of the modern thermal-magnetic breaker commonly used in household load centers to this day.

Interconnection of multiple generator sources into an electrical grid required the development of circuit breakers with increasing voltage ratings and increased ability to safely interrupt the increasing short-circuit currents produced by networks. Simple air-break manual switches produced hazardous arcs when interrupting high voltages; these gave way to oil-enclosed contacts, and various forms using the directed flow of pressurized air, or of pressurized oil, to cool and interrupt the arc. By 1935, the specially constructed circuit breakers used at the Boulder Dam project use eight series breaks and pressurized oil flow to interrupt faults of up to 2,500 MVA, in three cycles of the AC power frequency.

Operation

All circuit breaker systems have common features in their operation, but details vary substantially depending on the voltage class, current rating and type of the circuit breaker.

The circuit breaker must first detect a fault condition. In small mains and low voltage circuit breakers, this is usually done within the device itself. Typically, the heating or magnetic effects of electric current are employed. Circuit breakers for large currents or high voltages are usually arranged with protective relay pilot devices to sense a fault condition and to operate the opening mechanism. These typically require a separate power source, such as a battery, although some high-voltage circuit breakers are self-contained with current transformers, protective relays, and an internal control power source.

Once a fault is detected, the circuit breaker contacts must open to interrupt the circuit; this is commonly done using mechanically stored energy contained within the breaker, such as a spring or compressed air to separate the contacts. Circuit breakers may also use the higher current caused by the fault to separate the contacts, such as thermal expansion or a magnetic field. Small circuit breakers typically have a manual control lever to switch off the load or reset a tripped breaker, while larger units use solenoids to trip the mechanism, and electric motors to restore energy to the springs.

The circuit breaker contacts must carry the load current without excessive heating, and must also withstand the heat of the arc produced when interrupting (opening) the circuit. Contacts are made of copper or copper alloys, silver alloys and other highly conductive materials. Service life of the contacts is limited by the erosion of contact material due to arcing while interrupting the current. Miniature and molded-case circuit breakers are usually discarded when the contacts have worn, but power circuit breakers and high-voltage circuit breakers have replaceable contacts.

When a high current or voltage is interrupted, an arc is generated. The length of the arc is generally proportional to the voltage while the intensity (or heat) is proportional to the current. This arc must be contained, cooled and extinguished in a controlled way, so that the gap between the contacts can again withstand the voltage in the circuit. Different circuit breakers use vacuum, air, insulating gas, or oil as the medium the arc forms in. Different techniques are used to extinguish the arc including:

- Lengthening or deflecting the arc.
- Intensive cooling (in jet chambers).
- Division into partial arcs.
- Zero point quenching (contacts open at the zero current time crossing of the AC waveform, effectively breaking no load current at the time of opening. The zero crossing occurs at twice the line frequency; i.e., 100 times per second for 50 Hz and 120 times per second for 60 Hz AC.)
- Connecting capacitors in parallel with contacts in DC circuits.

Finally, once the fault condition has been cleared, the contacts must again be closed to restore power to the interrupted circuit.

Arc Interruption

Low-voltage miniature circuit breakers (MCB) use air alone to extinguish the arc. These circuit breakers contain so-called arc chutes, a stack of mutually insulated parallel metal plates that divide and cool the arc. By splitting the arc into smaller arcs the arc is cooled down while the arc voltage is increased and serves as an additional impedance that limits the current through the circuit breaker. The current-carrying parts near the contacts provide easy deflection of the arc into the arc chutes by a magnetic force of a current path, although magnetic blowout coils or permanent magnets could also deflect the arc into the arc chute (used on circuit breakers for higher ratings). The number of plates in the arc chute is dependent on the short-circuit rating and nominal voltage of the circuit breaker.

In larger ratings, oil circuit breakers rely upon vaporization of some of the oil to blast a jet of oil through the arc.

Gas (usually sulfur hexafluoride) circuit breakers sometimes stretch the arc using a magnetic field, and then rely upon the dielectric strength of the sulfur hexafluoride (SF_6) to quench the stretched arc.

Vacuum circuit breakers have minimal arcing (as there is nothing to ionize other than the contact material). The arc quenches when it is stretched a very small amount (less than 2–3 mm (0.08–0.1 in)). Vacuum circuit breakers are frequently used in modern medium-voltage switchgear to 38,000 volts.

Air circuit breakers may use compressed air to blow out the arc, or alternatively, the contacts are rapidly swung into a small sealed chamber, the escaping of the displaced air thus blowing out the arc.

Circuit breakers are usually able to terminate all current very quickly: typically the arc is extinguished between 30 ms and 150 ms after the mechanism has been tripped, depending upon age and construction of the device. The maximum current value and let-through energy determine the quality of the circuit breakers.

Short-circuit

Circuit breakers are rated both by the normal current that they are expected to carry, and the maximum short-circuit current that they can safely interrupt. This latter figure is the ampere interrupting capacity (AIC) of the breaker.

Under short-circuit conditions, the calculated or measured maximum prospective short-circuit current may be many times the normal, rated current of the circuit. When electrical contacts open to interrupt a large current, there is a tendency for an arc to form between the opened contacts, which would allow the current to continue. This condition can create conductive ionized gases and molten or vaporized metal, which can cause further continuation of the arc, or creation of additional short circuits, potentially resulting in the explosion of the circuit breaker and the equipment that it is installed in. Therefore, circuit breakers must incorporate various features to divide and extinguish the arc.

The maximum short-circuit current that a breaker can interrupt is determined by testing. Application of a breaker in a circuit with a prospective short-circuit current higher than the breaker's interrupting capacity rating may result in failure of the breaker to safely interrupt a fault. In a worst-case scenario the breaker may successfully interrupt the fault, only to explode when reset.

Typical domestic panel circuit breakers are rated to interrupt 10 kA (10000 A) short-circuit current.

Miniature circuit breakers used to protect control circuits or small appliances may not have sufficient interrupting capacity to use at a panel board; these circuit breakers are called "supplemental circuit protectors" to distinguish them from distribution-type circuit breakers.

Standard Current Ratings

Circuit breakers are manufactured in standard sizes, using a system of preferred numbers to cover a range of ratings. Miniature circuit breakers have a fixed trip setting; changing the operating current value requires changing the whole circuit breaker. Larger circuit breakers can have adjustable trip settings, allowing standardized elements to be applied but with a setting intended to improve protection. For example, a circuit breaker with a 400 ampere "frame size" might have its overcurrent detection set to operate at only 300 amperes, to protect a feeder cable.

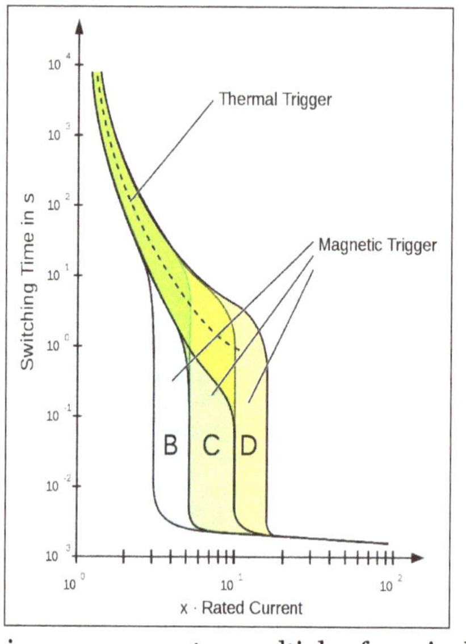

Time till trip versus current as multiple of nominal current.

International Standards, IEC 60898-1 and European Standard EN 60898-1, define the *rated current* I_n of a circuit breaker for low voltage distribution applications as the maximum current that the breaker is designed to carry continuously (at an ambient air temperature of 30 °C). The commonly available preferred values for the rated current are 1 A, 2 A, 4 A, 6 A, 10 A, 13 A, 16 A, 20 A, 25 A, 32 A, 40 A, 50 A, 63 A, 80 A, 100 A, and 125 A (similar to the R10 Renard series, but using 6, 13, and 32 instead of 6.3, 12.5, and 31.5 – it includes the 13 A current limit of British BS 1363 sockets). The circuit breaker is labeled with the rated current in amperes, but excluding the unit symbol, A. Instead, the ampere figure is preceded by a letter, *B*, *C*, or *D*, which indicates the *instantaneous tripping current* — that is, the minimum value of current that causes the circuit breaker to trip without intentional time delay (i.e., in less than 100 ms), expressed in terms of I_n:

Type	Instantaneous tripping current
B	Above 3 I_n
C	Above 5 I_n up to and including 10 I_n
D	Above 10 I_n up to and including 20 I_n
K	Above 8 I_n up to and including 12 I_n For the protection of loads that cause frequent short duration (approximately 400 ms to 2 s) current peaks in normal operation.
Z	Above 2 I_n up to and including 3 I_n for periods in the order of tens of seconds. For the protection of loads such as semiconductor devices or measuring circuits using current transformers.

Circuit breakers are also rated by the maximum fault current that they can interrupt; this allows use of more economical devices on systems unlikely to develop the high short-circuit current found on, for example, a large commercial building distribution system.

In the United States, Underwriters Laboratories (UL) certifies equipment ratings, called Series Ratings (or "integrated equipment ratings") for circuit breaker equipment used for buildings.

Power circuit breakers and medium- and high-voltage circuit breakers used for industrial or electric power systems are designed and tested to ANSI or IEEE standards in the C37 series.

Types

Front panel of a 1250 A air circuit breaker manufactured by ABB. This low-voltage power circuit breaker can be withdrawn from its housing for servicing. Trip characteristics are configurable via DIP switches on the front panel.

Many classifications of circuit breakers can be made, based on their features such as voltage class, construction type, interrupting type, and structural features.

Low-voltage

Low-voltage (less than 1,000 V_{AC}) types are common in domestic, commercial and industrial application, and include:

- Miniature circuit breaker (MCB)—rated current not more than 100 A. Trip characteristics normally not adjustable. Thermal or thermal-magnetic operation. Breakers illustrated above are in this category.
- Molded Case Circuit Breaker (MCCB)—rated current up to 2,500 A. Thermal or thermal-magnetic operation. Trip current may be adjustable in larger ratings.
- Low-voltage power circuit breakers can be mounted in multi-tiers in low-voltage switchboards or switchgear cabinets.

The characteristics of low-voltage circuit breakers are given by international standards such as IEC 947. These circuit breakers are often installed in draw-out enclosures that allow removal and interchange without dismantling the switchgear.

Large low-voltage molded case and power circuit breakers may have electric motor operators so they can open and close under remote control. These may form part of an automatic transfer switch system for standby power.

Low-voltage circuit breakers are also made for direct-current (DC) applications, such as DC for subway lines. Direct current requires special breakers because the arc is continuous—unlike an AC arc, which tends to go out on each half cycle. A direct current circuit breaker has blow-out coils that generate a magnetic field that rapidly stretches the arc. Small circuit breakers are either installed directly in equipment, or are arranged in a breaker panel.

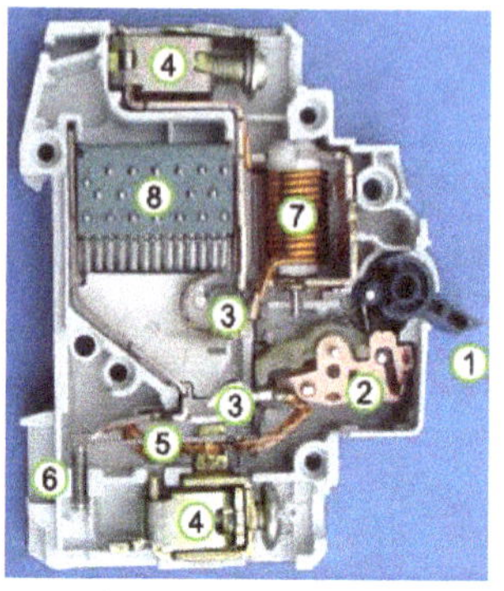

Inside of a circuit breaker.

The DIN rail-mounted thermal-magnetic miniature circuit breaker is the most common style in modern domestic consumer units and commercial electrical distribution boards throughout Europe. The design includes the following components:

- Actuator lever - used to manually trip and reset the circuit breaker. Also indicates the status of the circuit breaker (On or Off/tripped). Most breakers are designed so they can still trip even if the lever is held or locked in the "on" position. This is sometimes referred to as "free trip" or "positive trip" operation.
- Actuator mechanism - forces the contacts together or apart.
- Contacts - allow current when touching and break the current when moved apart.
- Terminals.
- Bimetallic strip - separates contacts in response to smaller, longer-term overcurrents.
- Calibration screw - allows the manufacturer to precisely adjust the trip current of the device after assembly.
- Solenoid - separates contacts rapidly in response to high overcurrents.
- Arc divider/extinguisher.

Solid State

Solid-state circuit breakers, also known as *digital circuit breakers* are a technological innovation which promises advance circuit breaker technology out of the mechanical level, into the electrical. This promises several advantages, such as cutting the circuit in fractions of microseconds, better monitoring of circuit loads and longer lifetimes.

Magnetic

Magnetic circuit breakers use a solenoid (electromagnet) whose pulling force increases with the current. Certain designs utilize electromagnetic forces in addition to those of the solenoid. The circuit breaker contacts are held closed by a latch. As the current in the solenoid increases beyond the rating of the circuit breaker, the solenoid's pull releases the latch, which lets the contacts open by spring action. They are the most commonly used circuit breakers in the USA.

Thermal-magnetic

Shihlin Electric MCCB with SHT.

Thermal magnetic circuit breakers, which are the type found in most distribution boards in Europe and countries with a similar wiring arrangements, incorporate both techniques with the electromagnet responding instantaneously to large surges in current (short circuits) and the bimetallic strip responding to less extreme but longer-term over-current conditions. The thermal portion of the circuit breaker provides a time response feature, that trips the circuit breaker sooner for larger over currents but allows smaller overloads to persist for a longer time. This allows short current spikes such as are produced when a motor or other non-resistive load is switched on. With very large over-currents during a short-circuit, the magnetic element trips the circuit breaker with no intentional additional delay.

Magnetic-hydraulic

A *magnetic-hydraulic* circuit breaker uses a solenoid coil to provide operating force to open the contacts. Magnetic-hydraulic breakers incorporate a hydraulic time delay feature using a viscous fluid. A spring restrains the core until the current exceeds the breaker rating. During an overload, the speed of the solenoid motion is restricted by the fluid. The delay permits brief current surges beyond normal running current for motor starting, energizing equipment, etc. Short-circuit currents provide sufficient solenoid force to release the latch regardless of core position thus bypassing the delay feature. Ambient temperature affects the time delay but does not affect the current rating of a magnetic breaker.

Large power circuit breakers, applied in circuits of more than 1000 volts, may incorporate hydraulic elements in the contact operating mechanism. Hydraulic energy may be supplied by a pump, or stored in accumulators. These form a distinct type from oil-filled circuit breakers where oil is the arc extinguishing medium.

Common Trip (Ganged) Breakers

To provide simultaneous breaking on multiple circuits from a fault on any one, circuit breakers may be made as a ganged assembly. This is a very common requirement for 3 phase systems, where breaking may be either 3 or 4 pole (solid or switched neutral). Some makers make ganging kits to allow groups of single phase breakers to be interlinked as required.

Three-pole common trip breaker for supplying a three-phase device. This breaker has a 2 A rating.

In the US, where split phase supplies are common, in branch circuits with more than one live conductor, each live conductor must be protected by a breaker pole. To ensure that all live conductors are interrupted when any pole trips, a "common trip" breaker must be used. These may either contain two or three tripping mechanisms within one case, or for small breakers, may externally tie the poles together via their operating handles. Two-pole common trip breakers are common on 120/240-volt systems where 240 volt loads (including major appliances or further distribution boards) span the two live wires. Three-pole common trip breakers are typically used to supply three-phase electric power to large motors or further distribution boards.

Separate circuit breakers must never be used for live and neutral, because if the neutral is disconnected while the live conductor stays connected, a very dangerous condition arises: the circuit appears de-energized (appliances don't work), but wires remain live and some residual-current devices (RCDs) may not trip if someone touches the live wire (because some RCDs need power to trip). This is why only common trip breakers must be used when neutral wire switching is needed.

Shunt-trip units

A shunt-trip unit appears similar to a normal breaker and the moving actuators are 'ganged' to a normal breaker mechanism to operate together in a similar way, but the shunt trip is a solenoid intended to be operated by an external constant voltage signal, rather than a current, commonly the local mains voltage or DC. These are often used to cut the power when a high risk event occurs, such as a fire or flood alarm, or another electrical condition, such as over voltage detection. Shunt trips may be a user fitted accessory to a standard breaker, or supplied as an integral part of the circuit breaker.

Medium-voltage

Medium-voltage circuit breakers rated between 1 and 72 kV may be assembled into metal-enclosed switchgear line ups for indoor use, or may be individual components installed outdoors in a substation. Air-break circuit breakers replaced oil-filled units for indoor applications, but are now themselves being replaced by vacuum circuit breakers (up to about 40.5 kV). Like the high voltage circuit breakers described below, these are also operated by current sensing protective relays operated through current transformers. The characteristics of MV breakers are given by international standards such as IEC 62271. Medium-voltage circuit breakers nearly always use separate current sensors and protective relays, instead of relying on built-in thermal or magnetic overcurrent sensors.

An air circuit breaker of Siemens brand mounted on a motor control cubicle.

Medium-voltage circuit breakers can be classified by the medium used to extinguish the arc:

- Vacuum circuit breakers—With rated current up to 6,300 A, and higher for generator circuit breakers application (up to 16,000 A & 140 kA). These breakers interrupt the current by creating and extinguishing the arc in a vacuum container - aka "bottle". Long life bellows are designed to travel the 6–10 mm the contacts must part. These are generally applied for voltages up to about 40,500 V, which corresponds roughly to the medium-voltage range of power systems. Vacuum circuit breakers have longer life expectancy between overhaul than do other circuit breakers. In addition their global warming potential is by far lower than SF_6 circuit breaker.
- Air circuit breakers—Rated current up to 6,300 A and higher for generator circuit breakers. Trip characteristics are often fully adjustable including configurable trip thresholds and delays. Usually electronically controlled, though some models are microprocessor controlled via an integral electronic trip unit. Often used for main power distribution in large industrial plant, where the breakers are arranged in draw-out enclosures for ease of maintenance.
- SF_6 circuit breakers extinguish the arc in a chamber filled with sulfur hexafluoride gas.

Medium-voltage circuit breakers may be connected into the circuit by bolted connections to bus bars or wires, especially in outdoor switchyards. Medium-voltage circuit breakers in switchgear line-ups are often built with draw-out construction, allowing breaker removal without disturbing power circuit connections, using a motor-operated or hand-cranked mechanism to separate the breaker from its enclosure.

High-voltage

Electrical power transmission networks are protected and controlled by high-voltage breakers. The definition of *high voltage* varies but in power transmission work is usually thought to be 72.5 kV or higher, according to a recent definition by the International Electrotechnical Commission (IEC). High-voltage breakers are nearly always solenoid-operated, with current sensing protective relays operated through current transformers. In substations the protective relay scheme can be complex, protecting equipment and buses from various types of overload or ground/earth fault.

Three single-phase Soviet/Russian 110-kV oil circuit breakers.

400 kV SF_6 live-tank circuit breakers.

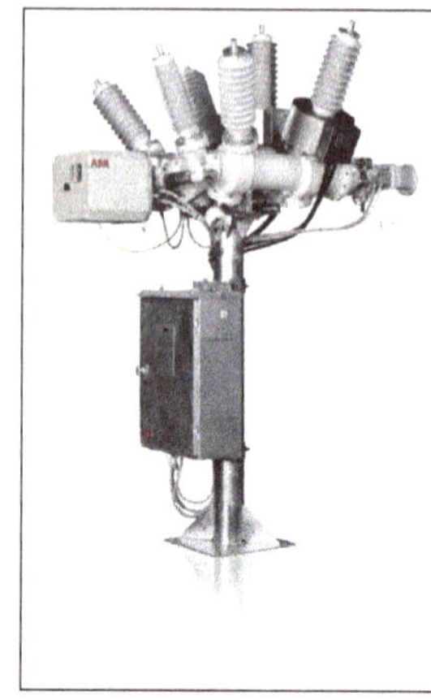

72.5 kV hybrid switchgear module.

High-voltage breakers are broadly classified by the medium used to extinguish the arc:

- Bulk oil
- Minimum oil
- Air blast
- Vacuum
- SF_6
- CO_2

Due to environmental and cost concerns over insulating oil spills, most new breakers use SF_6 gas to quench the arc.

Circuit breakers can be classified as *live tank*, where the enclosure that contains the breaking mechanism is at line potential, or *dead tank* with the enclosure at earth potential. High-voltage AC circuit breakers are routinely available with ratings up to 765 kV. 1,200 kV breakers were launched by Siemens in November 2011, followed by ABB in April the following year.

High-voltage circuit breakers used on transmission systems may be arranged to allow a single pole of a three-phase line to trip, instead of tripping all three poles; for some classes of faults this improves the system stability and availability.

High-voltage direct current circuit breakers are still a field of research as of 2015. Such breakers would be useful to interconnect HVDC transmission systems.

Sulfur Hexafluoride (SF6) High-voltage

A sulfur hexafluoride circuit breaker uses contacts surrounded by sulfur hexafluoride gas to quench the arc. They are most often used for transmission-level voltages and may be incorporated into compact gas-insulated switchgear. In cold climates, supplemental heating or de-rating of the circuit breakers may be required due to liquefaction of the SF6 gas.

Disconnecting Circuit Breaker (DCB)

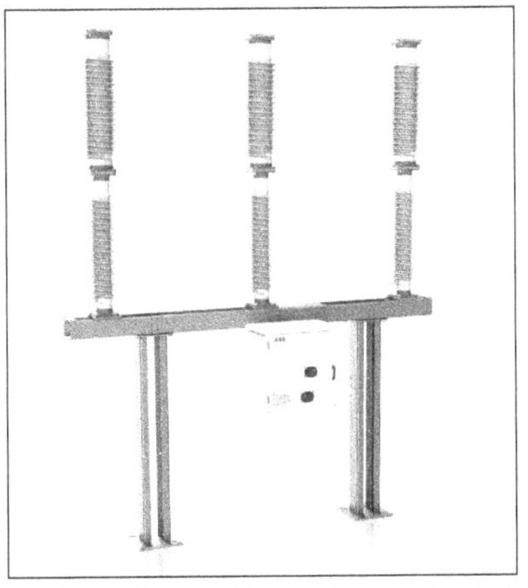

72.5 kV carbon dioxide high-voltage circuit breaker.

The disconnecting circuit breaker (DCB) was introduced in 2000 and is a high-voltage circuit breaker modeled after the SF_6-breaker. It presents a technical solution where the disconnecting function is integrated in the breaking chamber, eliminating the need for separate disconnectors. This increases the availability, since open-air disconnecting switch main contacts need maintenance every 2–6 years, while modern circuit breakers have maintenance intervals of 15 years. Implementing a DCB solution also reduces the space requirements within the substation, and increases the reliability, due to the lack of separate disconnectors.

In order to further reduce the required space of substation, as well as simplifying the design and engineering of the substation, a fiber optic current sensor (FOCS) can be integrated with the DCB. A 420 kV DCB with integrated FOCS can reduce a substation's footprint with over 50% compared to a conventional solution of live tank breakers with disconnectors and current transformers, due to reduced material and no additional insulation medium.

Carbondioxide (CO_2) High-voltage

In 2012 ABB presented a 75 kV high-voltage breaker that uses carbon dioxide as the medium to extinguish the arc. The carbon dioxide breaker works on the same principles as an SF_6 breaker and can also be produced as a disconnecting circuit breaker. By switching from SF_6 to CO_2 it is possible to reduce the CO_2 emissions by 10 tons during the product's life cycle.

Smart Circuit Breakers

Several firms have looked at adding monitoring for appliances via electronics or using a digital circuit breaker to monitor the breakers remotely. Utility companies in the United States have been reviewing use of the technology to turn on and off appliances, as well as potentially turning off charging of electric cars during periods of high electrical grid load. These devices under research and testing would have wireless capability to monitor the electrical usage in a house via a smartphone app or other means.

Other Breakers

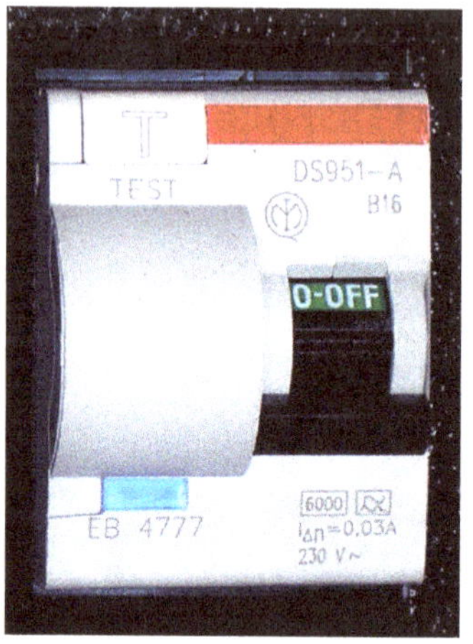

Residual current circuit breaker with overload protection.

The following types are described in separate articles.

- Breakers for protections against earth faults too small to trip an over-current device:
    - RCCB (Residual Current Operated Circuit-Breaker without Integral Overcurrent Protection) as a type of Residual-current device (RCD, formerly known as a *residual current device*) — detects current imbalance, but does not provide over-current protection. In the United States and Canada these are called *ground fault circuit interrupters (GFCI)*.
    - Residual current breaker with over-current protection (RCBO) — combines the functions of an RCD and an MCB in one package. In the United States and Canada these are called GFCI breakers.
    - Earth leakage circuit breaker (ELCB) — This detects current in the earth wire directly rather than detecting imbalance. They are no longer seen in new installations as they cannot detect any dangerous condition where the current is returning to earth by another route - such as via a person on the ground or via plumbing. (also called VOELCB in the UK).
- Recloser — A type of circuit breaker that closes automatically after a delay. These are used on overhead electric power distribution systems, to prevent short duration faults from causing sustained outages.
- Polyswitch (polyfuse) — A small device commonly described as an automatically resetting fuse rather than a circuit breaker.

Air Circuit Breaker

This type of circuit breakers, is those kind of circuit breaker which operates in air at atmospheric pressure. After development of oil circuit breaker, the medium voltage air circuit breaker (ACB) is replaced completely by oil circuit breaker in different countries. But in countries like France and Italy, ACBs are still preferable choice up to voltage 15 KV. It is also good choice to avoid the risk of oil fire, in case of oil circuit breaker. In America ACBs were exclusively used for the system up to 15 KV until the development of new vacuum and SF6 circuit breakers.

Working Principle of Air Circuit Breaker

The working principle of this breaker is rather different from those in any other types of circuit breakers. The main aim of all kind of circuit breaker is to prevent the reestablishment of arcing after current zero by creating a situation where in the contact gap will withstand the system recovery voltage. The air circuit breaker does the same but in different manner. For interrupting arc it creates an arc voltage in excess of the supply voltage. Arc voltage is defined as the minimum voltage required maintaining the arc. This circuit breaker increases the arc voltage by mainly three different ways:

- It may increase the arc voltage by cooling the arc plasma. As the temperature of arc plasma is decreased, the mobility of the particle in arc plasma is reduced; hence more voltage gradient is required to maintain the arc.
- It may increase the arc voltage by lengthening the arc path. As the length of arc path is increased, the resistance of the path is increased, and hence to maintain the same arc current more voltage is required to be applied across the arc path. That means arc voltage is increased.
- Splitting up the arc into a number of series arcs also increases the arc voltage.

Types of ACB

There are mainly two types of ACB are available:

- Plain air circuit breaker.
- Air blast Circuit Breaker.

Operation of ACB

- The first objective is usually achieved by forcing the arc into contact with as large an area as possible of insulating material. Every air circuit breaker is fitted with a chamber surrounding the contact. This chamber is called 'arc chute'. The arc is driven into it. If inside of the arc chute is suitably shaped, and if the arc can be made conform to the shape, the arc chute wall will help to achieve cooling. This type of arc chute should be made from some kind of refractory material. High temperature plastics reinforced with glass fiber and ceramics are preferable materials for making arc chute.
- The second objective that is lengthening the arc path, is achieved concurrently with fist objective. If the inner walls of the arc chute is shaped in such a way that the arc is not only forced into close proximity with it but also driven into a serpentine channel projected on the arc chute wall. The lengthening of the arc path increases the arc resistance.
- The third technique is achieved by using metal arc slitter inside the arc chute. The main arc chute is divided into numbers of small compartments by using metallic separation plates.

These metallic separation plates are actually the arc splitters and each of the small compartments behaves as individual mini arc chute. In this system the initial arc is split into a number of series arcs, each of which will have its own mini arc chute. So each of the split arcs has its own cooling and lengthening effect due to its own mini arc chute and hence individual split arc voltage becomes high. These collectively, make the overall arc voltage, much higher than the system voltage.

The air circuit breaker, operated within the voltage level 1 KV, does not require any arc control device. Mainly for heavy fault current on low voltages (low voltage level above 1 KV) ABCs with appropriate arc control device, are good choice. These breakers normally have two pairs of contacts. The main pair of contacts carries the current at normal load and these contacts are made of copper. The additional pair is the arcing contact and is made of carbon. When circuit breaker is being opened, the main contacts open first and during opening of main contacts the arcing contacts are still in touch with each other. As the current gets, a parallel low resistive path through the arcing contact during opening of main contacts, there will not be any arcing in the main contact. The arcing is only initiated when finally the arcing contacts are separated. The each of the arc contacts is fitted with an arc runner which helps, the arc discharge to move upward due to both thermal and electromagnetic effects as shown in the figure. As the arc is driven upward it enters in the arc chute, consisting of splitters. The arc in chute will become colder, lengthen and split hence arc voltage becomes much larger than system voltage at the time of operation of air circuit breaker, and therefore the arc is quenched finally during the current zero.

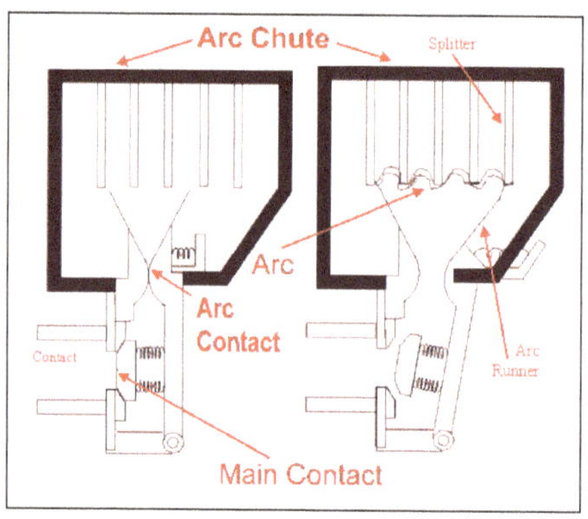

Although these types of circuit breakers have become obsolete for medium voltage application, but they are still preferable choice for high current rating in low voltage application.

Air Blast Circuit Breaker

These types of air circuit breaker were used for the system voltage of 245 KV, 420 KV and even more, especially where faster breaker operation was required. Air blast circuit breaker has some specific advantages over oil circuit breaker which are listed as follows:

- There is no chance of fire hazard caused by oil.
- The breaking speed of circuit breaker is much higher during operation of air blast circuit breaker.
- Arc quenching is much faster during operation of air blast circuit breaker.

- The duration of arc is same for all values of small as well as high currents interruptions.
- As the duration of arc is smaller, so lesser amount of heat realized from arc to current carrying contacts hence the service life of the contacts becomes longer.
- The stability of the system can be well maintained as it depends on the speed of operation of circuit breaker.
- Requires much less maintenance compared to oil circuit breaker.

There are also some disadvantages of air blast circuit breakers:

- In order to have frequent operations, it is necessary to have sufficiently high capacity air compressor.
- Frequent maintenance of compressor, associated air pipes and automatic control equipments is also required.
- Due to high speed current interruption there is always a chance of high rate of rise of re-striking voltage and current chopping.
- There also a chance of air pressure leakage from air pipes junctions.

There are mainly two types of ACB, plain air circuit breaker and air blast circuit breaker. But the later can be sub divided further into three different categories.

- Axial Blast ACB.
- Axial Blast ACB with side moving contact.
- Cross Blast ACB.

Axial Blast Air Circuit Breaker

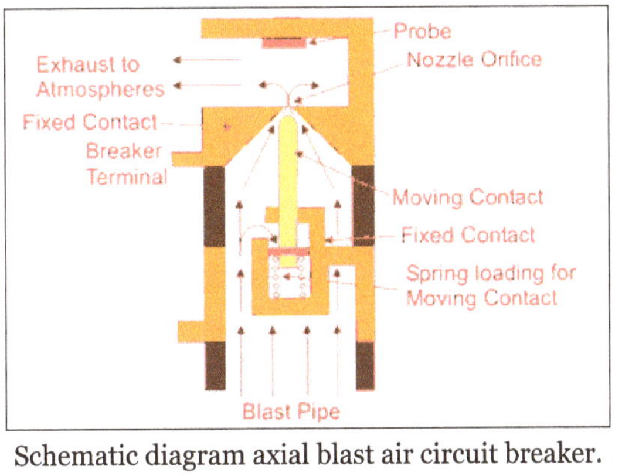

Schematic diagram axial blast air circuit breaker.

In axial blast ACB the moving contact is in contact with fixed contact with the help of a spring pressure as shown in the figure. There is a nozzle orifice in the fixed contact which is blocked by tip of the moving contact at normal closed condition of the breaker. When fault occurs, the high pressure air is introduced into the arcing chamber. The air pressure will counter the spring pressure and deforms the spring hence the moving contact is withdrawn from the fixed contact and nozzle hole becomes open. At the same time the high pressure air starts flowing along the arc through the fixed contact nozzle orifice. This axial flow of air along the arc through the nozzle orifice will make the arc lengthen and colder hence arc voltage become much higher than system voltage that means system voltage is insufficient to sustain the arc consequently the arc is quenched.

Axial Blast ACB with Side Moving Contact

In this type of axial blast air circuit breaker the moving contact is fitted over a piston supported over a spring. In order to open the circuit breakerthe air is admitted into the arcing chamber when pressure reaches to a predetermined value, it presses down the moving contact; an arc is drawn between the fixed and moving contacts. The air blast immediately transfers the arc to the arcing electrode and is consequently quenched by the axial flow of air.

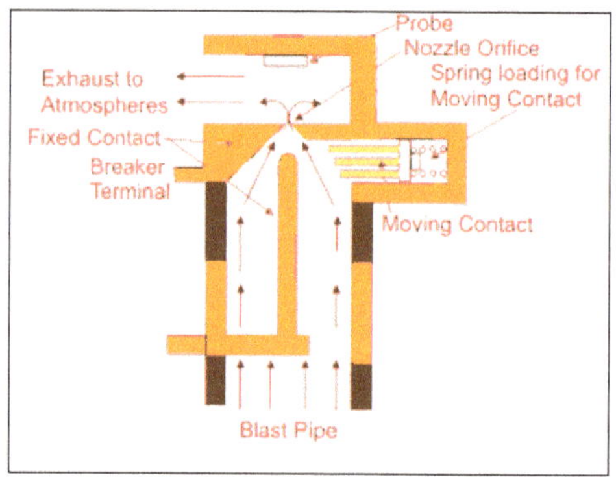

Axial blast air circuit breaker with side moving contact.

Cross Blast Air Circuit Breaker

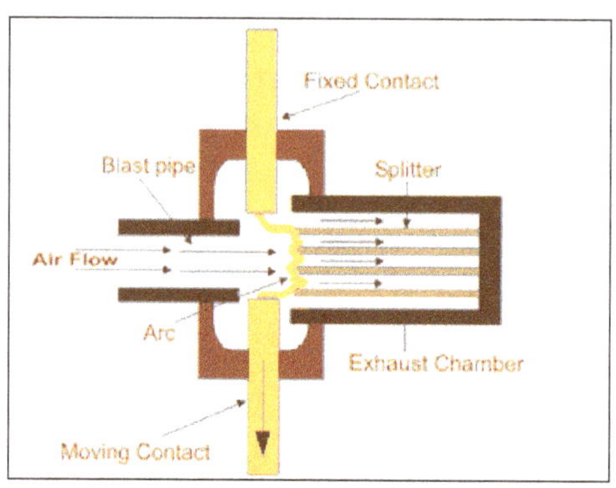

Principle cross blast air circuit breaker.

The working principle of cross blast air circuit breaker is quite simple. In this system of air blast circuit breaker the blast pipe is fixed in perpendicular to the movement of moving contact in the arcing chamber and on the opposite side of the arcing chamber one exhaust chamber is also fitted at the same alignment of blast pipe, so that the air comes from blast pipe can straightly enter into exhaust chamber through the contact gap of the breaker. The exhaust chamber is spit with arc splitters. When moving contact is withdrawn from fixed contact, an arc is established in between the contact, and at the same time high pressure air coming from blast pipe will pass through the contact gap and will forcefully take the arc into exhaust chamber where the arc is split with the help of arc splitters and ultimately arc is quenched.

SF6 Circuit Breaker

SF_6 gas has excellent insulating property and it has a high electronegativity. Due to high electronegativity, it has a high affinity for absorbing free electrons. It absorbes electrons to form a negative ions. The attachment of electron with SF_6 molecules may occur in two different reactions as shown below:

$$SF_6 + e = SF_6^-$$
$$SF_6 + e = SF_5^- + F$$

Ions are obviously much heavier than a free electron and therefore over all mobility of charged particles in SF_6 gas is much less as compared to the other common gases.

We know that the mobility of charged particles is majorly responsible for conducting current through a gas. Hence, for heavier and less mobile charged particles in SF_6 gas, it acquires very high dielectric strength.

400 KV SF6 Circuit Breaker.

Not only the gas has a good dielectric strength but also it has the unique property of fast recombination after the arc is removed. The gas also has excellent heat transfer property. Due to its low gaseous viscosity, SF6 gas can efficiently transfer heat by convection. So due to high dielectric strength and high cooling effect, SF_6 gas is approximately 100 times more effective arc quenching media than air. Due to these unique properties of this gas, SF6 circuit breaker is used in a complete range of medium voltage and high voltage electrical power system. These circuit breakersare available for the voltage ranges from 33KV to 800 KV and even more.

Disadvantages of SF6 Circuit Breaker

- The SF6 gas is identified as a greenhouse gas, safety regulation is being introduced in many countries in order to prevent its release into the atmosphere.
- Puffer type design of SF_6 CB needs a high mechanical energy which is almost five times greater than that of oil circuit breaker.

Types of SF_6 Circuit Breaker

There are mainly three types of SF6 CB depending upon the voltage level of application:

- Single interrupter SF6 CB applied for up to 245 KV(220 KV) system.

- Two interrupter SF6 CB applied for up to 420 KV(400 KV) system.
- Four interrupter SF6 CB applied for up to 800 KV(715 KV) system.

Working of SF6 Circuit Breaker

The working of SF6 CB of first generation was quite simple and it is some extent similar to air blast circuit breaker. Here SF6 gas was compressed and stored in a high-pressure reservoir. During operation of SF6 circuit breaker, this highly compressed gas is released through the arc in the breaker and collected to a relatively low-pressure reservoir and then it is pumped back to the high-pressure reservoir for reutilizing.

The working of SF6 circuit breaker is a little bit different in modern time. The innovation of puffer type design makes the operation of SF6 CB much easier. In buffer type design, the arc energy is utilized to develop pressure in the arcing chamber for arc quenching. Here the breaker is filled with SF_6gas at rated pressure. There are two fixed contacts fitted with a specific contact gap. A sliding cylinder bridges these to fixed contacts. The cylinder can axially slide upward and downward along the contacts. There is one stationary piston inside the cylinder which is fixed with other stationary parts of the SF_6 circuit breaker, in such a way that it cannot change its position during the movement of the cylinder. As the piston is fixed and the cylinder is movable or sliding, the internal volume of the cylinder changes when the cylinder slides.

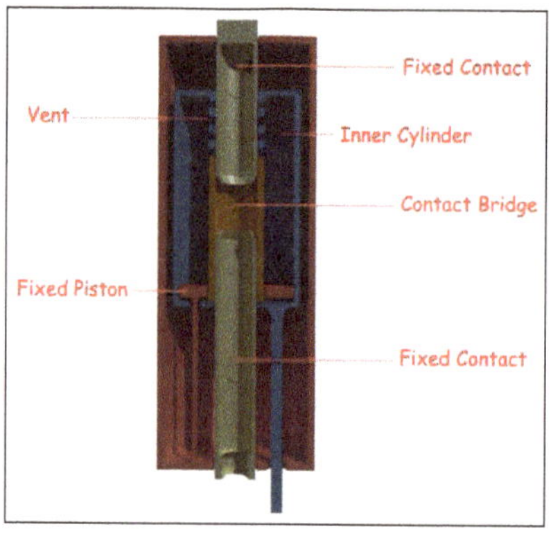

During the opening of the breaker the cylinder moves downwards against the position of the fixed piston hence the volume inside the cylinder is reduced which produces compressed SF6 gas inside the cylinder. The cylinder has numbers of side vents which were blocked by upper fixed contact body during closed position.

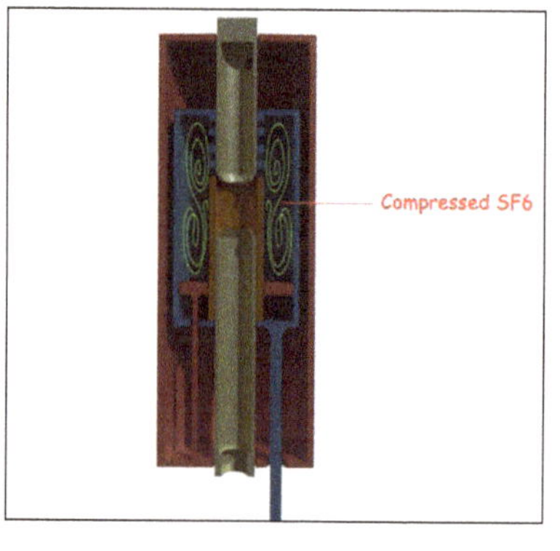

As the cylinder move further downwards, these vent openings cross the upper fixed contact, and become unblocked and then compressed SF6 gas inside the cylinder will come out through this vents in high speed towards the arc and passes through the axial hole of the both fixed contacts. The arc is quenched during this flow of SF6 gas.

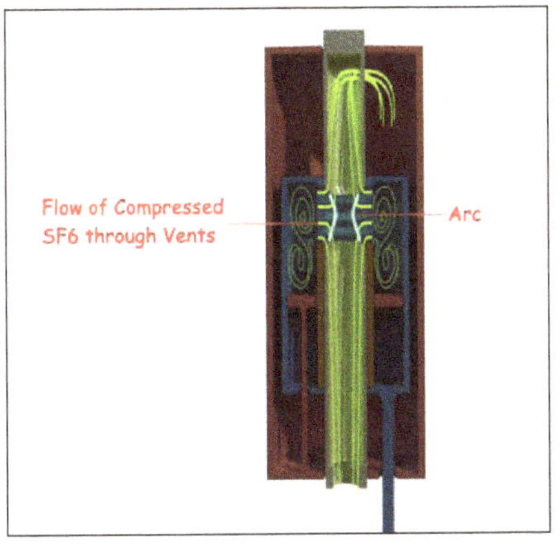

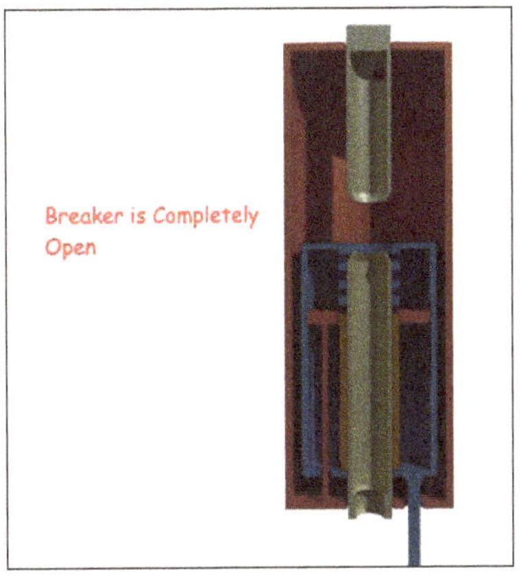

During closing of the circuit breaker, the sliding cylinder moves upwards and as the position of piston remains at a fixed height, the volume of the cylinder increases which introduces low pressure inside the cylinder compared to the surrounding. Due to this pressure difference, SF6 gas from surrounding will try to enter in the cylinder. The higher pressure gas will come through the axial hole of both fixed contact and enters into the cylinder via vent and during this flow; the gas will quench the arc.

References

- Electric-circuit, technology: britannica.com, Retrieved 22 May, 2019
- Huurdeman, Anton A. (2003). The Worldwide History of Telecommunications. USA: Wiley-IEEE. Pp. 199–200. ISBN 0-471-20505-2
- What-is-an-ac-circuit: circuitglobe.com, Retrieved 23 February, 2019
- Blanchard, Julian (October 1941). “The History of Electrical Resonance”. Bell System Technical Journal. USA: AT&T. 20 (4): 415. Doi:10.1002/j.1538-7305.1941.tb03608.x. Retrieved 2013-02-25
- Dc-circuit: circuitglobe.com, Retrieved 10 July, 2019
- Air-circuit-breaker-air-blast-circuit-breaker: electrical4u.com, Retrieved 13 March, 2019
- Types-and-operation-of-sf6-circuit-breaker: electrical4u.com, Retrieved 29 June, 2019

4

Electrical Laws and Circuit Theorems

Some of the major laws related to electrical circuits are Ohm's Law, Ampère's circuital law, Gauss's law, Coulomb's law, Faraday's laws of electromagnetic induction and Lenz's law. This chapter discusses in detail these laws related to electrical circuits as well as the different circuit theorems such as Superposition theorem, Thevenin's theorem, Norton's theorem, Maximum power transfer theorem etc.

Ohm's Law

Most basic components of electricity are voltage, current, and resistance. Ohm's law shows a simple relation between these three quantities. Ohm's law states that the current through a conductor between two points is directly proportional to the voltage across the two points.

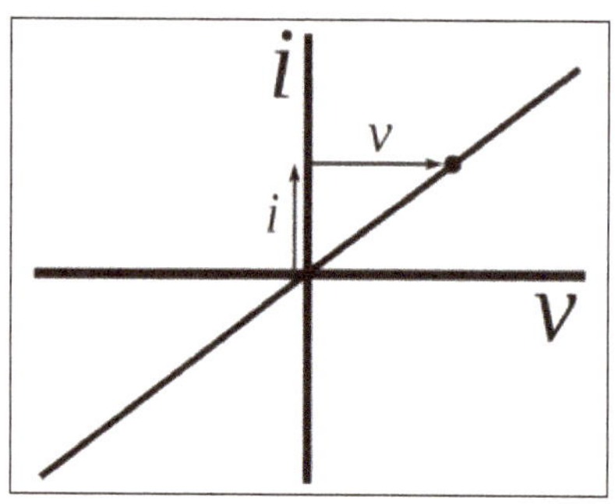

Ohm's Law Formula

Voltage= Current× Resistance

V= I×R

V= voltage, I= current and R= resistance

The SI unit of resistance is ohms and is denoted by Ω.

This law is one of the most basic laws of electricity. It helps to calculate the power, efficiency, current, voltage, and resistance of an element of an electrical circuit.

Applications of Ohm's Law

Ohm's law helps us in determining either voltage, current or impedance or resistance of a linear electric circuit when the other two quantities are known to us. It also makes power calculation simpler.

How do we Establish the Current-voltage Relationship?

In order to establish the current-voltage relationship, the ratio V / I remains constant for a given resistance, therefore a graph between the potential difference(V) and the current (I) must be a straight line.

How do we find the unknown values of resistance?

It is the constant ratio that gives the unknown values of resistance,

$$\frac{V}{I} = R$$

For a wire of uniform cross-section, the resistance depends on the length l and the area of cross-section A. It also depends on the temperature of the conductor. At a given temperature the resistance,

$$R = \rho \frac{1}{A}$$

where ρ is the specific resistance or resistivity and is characteristic of the material of wire. The specific resistance or resistivity of the material of the wire is,

$$\rho = \frac{RA}{1}$$

If 'r' is the radius of the wire, then the cross-sectional area, A = πr^2. Then the specific resistance or resistivity of the material of the wire is,

$$\rho = \frac{\pi r^2 R}{1}$$

Limitations of Ohms Law

1. Ohm's law is not applicable to unilateral networks. Unilateral networks allow the current to flow in one direction. Such types of network consist elements like a diode, transistor, etc.
2. Ohm's law is also not applicable to non – linear elements. Non-linear elements are those which do not have current exactly proportional to the applied voltage that means the resistance value of those elements changes for different values of voltage and current. Examples of non – linear elements are the thyristor.

Resistors

Resistors are one of the important blocks of electrical circuits. They are made up of the mixture of clay or carbon, so they are not only good conductors but good insulators too. Most of the resistors have four color bands. The first and second band reveal the first and second digits of the value respectively. The third band is used to multiply the value digits and the fourth band tells us the tolerance. If there is no fourth band, it is assumed that the tolerance is plus or minus 20%.

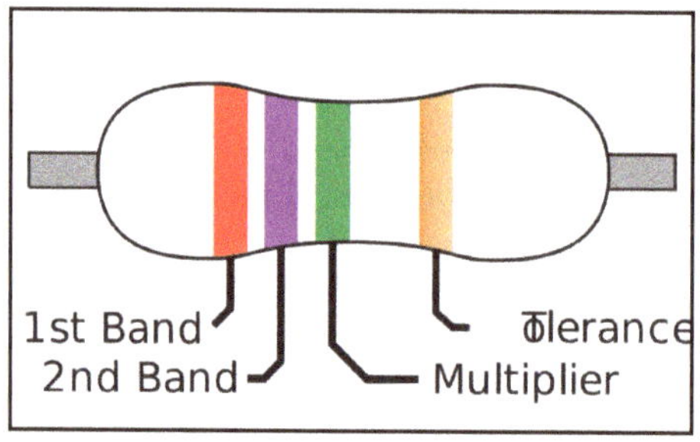

Resistance in Series

A series generally means connected along a line, or in a row, or in an order. In electronics, series resistance means that the resistors are connected one after the other and that there is only one path for current to flow through.

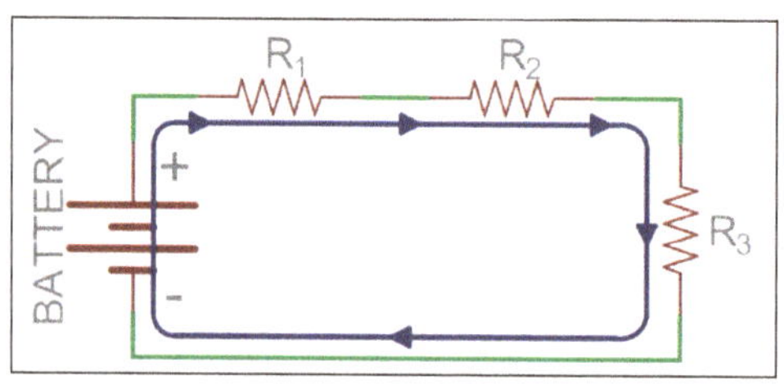

Laws of Series Circuits

- Individual resistance add up to the total circuit resistance.
- Current through the circuit is the same at every point.
- Individual voltages throughout the circuit add up to the total voltage.

Resistance in Parallel

There are many different ways to organize a parallel circuit. In the practical world, most of the wiring is done in parallel so that the voltage to any one part of the network is the same as the voltage supplied to any other part of it.

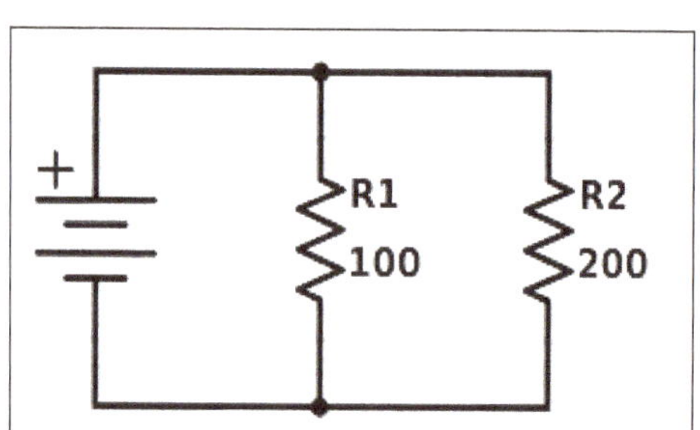

Laws of Parallel Circuits

- The reciprocals of all the individual resistances add up to the reciprocal of the total circuit resistance. 1/RT = 1/R1 + 1/R2 + 1/R3 ...
- Voltage through the circuit is the same at every point.
- Individual current draws throughout the circuit add up to the total current draw.

Ampère's Circuital Law

In classical electromagnetism, Ampère's circuital law relates the integrated magnetic field around a closed loop to the electric current passing through the loop. James Clerk Maxwell (not Ampère) derived it using hydrodynamics in his 1861 paper "On Physical Lines of Force" and it is now one of the Maxwell equations, which form the basis of classical electromagnetism.

The original form of Maxwell's circuital law, which he derived in his 1855 paper "On Faraday's Lines of Force" based on an analogy to hydrodynamics, relates magnetic fields to electric currents that produce them. It determines the magnetic field associated with a given current, or the current associated with a given magnetic field.

The original circuital law is only a correct law of physics in a magnetostatic situation, where the system is static except possibly for continuous steady currents within closed loops. For systems with electric fields that change over time, the original law must be modified to include a term known as Maxwell's correction.

Equivalent Forms

The original circuital law can be written in several different forms, which are all ultimately equivalent:

- An "integral form" and a "differential form". The forms are exactly equivalent, and related by the Kelvin–Stokes theorem.
- Forms using SI units, and those using cgs units. Other units are possible, but rare.
- Forms using either B or H magnetic fields. These two forms use the total current density and free current density, respectively. The B and H fields are related by the constitutive equation: $B = \mu_0 H$ where μ_0 is the magnetic constant.

Explanation

The integral form of the original circuital law is a line integral of the magnetic field around some closed curve *C* (arbitrary but must be closed). The curve *C* in turn bounds both a surface *S* which the electric current passes through (again arbitrary but not closed—since no three-dimensional volume is enclosed by *S*), and encloses the current. The mathematical statement of the law is a relation between the total amount of magnetic field around some path (line integral) due to the current which passes through that enclosed path (surface integral).

In terms of total current, (which is the sum of both free current and bound current) the line integral of the magnetic B-field (in teslas, T) around closed curve *C* is proportional to the total current I_{enc} passing through a surface *S* (enclosed by *C*). In terms of free current, the line integral of the magnetic H-field (in amperes per metre, A·m⁻¹) around closed curve *C* equals the free current $I_{f,enc}$ through a surface *S*.

Forms of the original circuital law written in SI units		
	Integral form	Differential form
Using B-field and total current	$\oint_C \mathbf{B}\cdot d\mathbf{l} = \mu_0 \iint_S \mathbf{J}\cdot d\mathbf{S} = \mu_0 I_{enc}$	$\nabla \times \mathbf{B} = \mu_0 \mathbf{J}$
Using H-field and free current	$\oint_C \mathbf{H}\cdot d\mathbf{l} = \iint_S \mathbf{J}_f\cdot d\mathbf{S} = I_{f,enc}$	$\nabla \times \mathbf{H} = \mathbf{J}_f$

- J is the total current density (in amperes per square metre, $A\cdot m^{-2}$),
- J_f is the free current density only,
- $\oint_C$ is the closed line integral around the closed curve *C*,
- $\iint_S$ denotes a 2-D surface integral over *S* enclosed by *C*,
- · is the vector dot product,
- d*l* is an infinitesimal element (a differential) of the curve *C* (i.e. a vector with magnitude equal to the length of the infinitesimal line element, and direction given by the tangent to the curve *C*),
- dS is the vector area of an infinitesimal element of surface *S* (that is, a vector with magnitude equal to the area of the infinitesimal surface element, and direction normal to surface *S*. The direction of the normal must correspond with the orientation of *C* by the right hand rule),
- $\nabla \times$ is the curl operator.

Ambiguities and Sign Conventions

There are a number of ambiguities in the above definitions that require clarification and a choice of convention.

- First, three of these terms are associated with sign ambiguities: the line integral $\oint_C$ could go around the loop in either direction (clockwise or counterclockwise); the vector area dS could point in either of the two directions normal to the surface; and I_{enc} is the net current passing through the surface *S*, meaning the current passing through in one direction, minus the current in the other direction—but either direction could be chosen as positive. These ambiguities are resolved by the right-hand rule: With the palm of the right-hand toward the area of integration, and the index-finger pointing along the direction of line-integration, the outstretched thumb points in the direction that must be chosen for the vector area dS. Also the current passing in the same direction as dS must be counted as positive. The right hand grip rule can also be used to determine the signs.

- Second, there are infinitely many possible surfaces *S* that have the curve *C* as their border. (Imagine a soap film on a wire loop, which can be deformed by moving the wire). Which of those surfaces is to be chosen? If the loop does not lie in a single plane, for example, there is no one obvious choice. The answer is that it does not matter; by Stoke's theorem, the integral is the same for any surface with boundary *C*, since the integrand is the curl of a smooth field (i.e. exact). In practice, one usually chooses the most convenient surface (with the given boundary) to integrate over.

Free Current versus Bound Current

The electric current that arises in the simplest textbook situations would be classified as "free current"—for example, the current that passes through a wire or battery. In contrast, "bound current" arises in the context of bulk materials that can be magnetized and/or polarized. (All materials can to some extent.)

When a material is magnetized (for example, by placing it in an external magnetic field), the electrons remain bound to their respective atoms, but behave as if they were orbiting the nucleus in a particular direction, creating a microscopic current. When the currents from all these atoms are put together, they create the same effect as a macroscopic current, circulating perpetually around the magnetized object. This magnetization current J_M is one contribution to "bound current".

The other source of bound current is bound charge. When an electric field is applied, the positive and negative bound charges can separate over atomic distances in polarizable materials, and when the bound charges move, the polarization changes, creating another contribution to the "bound current", the polarization current J_P.

The total current density J due to free and bound charges is then:

$$\mathbf{J} = \mathbf{J}_f + \mathbf{J}_M + \mathbf{J}_P,$$

with J_f the "free" or "conduction" current density.

All current is fundamentally the same, microscopically. Nevertheless, there are often practical reasons for wanting to treat bound current differently from free current. For example, the bound current usually originates over atomic dimensions, and one may wish to take advantage of a simpler theory intended for larger dimensions. The result is that the more microscopic Ampère's circuital law, expressed in terms of B and the microscopic current (which includes free, magnetization and polarization currents), is sometimes put into the equivalent form below in terms of H and the free current only.

Shortcomings of the Original Formulation of the Circuital Law

There are two important issues regarding the circuital law that require closer scrutiny. First, there is an issue regarding the continuity equation for electrical charge. In vector calculus, the identity for the divergence of a curl states that the divergence of the curl of a vector field must always be zero. Hence,

$$\nabla\cdot(\nabla\times\mathbf{B})=0,$$

and so the original Ampère's circuital law implies that,

$$\nabla\cdot\mathbf{J}=0.$$

But in general, reality follows the continuity equation for electric charge:

$$\nabla\cdot\mathbf{J}=-\frac{\partial\rho}{\partial t},$$

which is nonzero for a time-varying charge density. An example occurs in a capacitor circuit where time-varying charge densities exist on the plates.

Second, there is an issue regarding the propagation of electromagnetic waves. For example, in free space, where,

$$\mathbf{J}=\mathbf{0}.$$

The circuital law implies that,

$$\nabla\times\mathbf{B}=\mathbf{0},$$

but to maintain consistency with the continuity equation for electric charge, we must have,

$$\nabla\times\mathbf{B}=\frac{1}{c^2}\frac{\partial\mathbf{E}}{\partial t}.$$

To treat these situations, the contribution of displacement current must be added to the current term in the circuital law.

James Clerk Maxwell conceived of displacement current as a polarization current in the dielectric vortex sea, which he used to model the magnetic field hydrodynamically and mechanically. He added this displacement current to Ampère's circuital law at equation 112 in his 1861 paper "On Physical Lines of Force".

Displacement Current

In free space, the displacement current is related to the time rate of change of electric field. In a dielectric the above contribution to displacement current is present too, but a major contribution to the displacement current is related to the polarization of the individual molecules of the dielectric material. Even though charges cannot flow freely in a dielectric, the charges in molecules can move a little under the influence of an electric field. The positive and negative charges in molecules separate under the applied field, causing an increase in the state of polarization, expressed as the polarization density P. A changing state of polarization is equivalent to a current.

Both contributions to the displacement current are combined by defining the displacement current as:

$$\mathbf{J}_{\mathrm{D}} = \frac{\partial}{\partial t}\mathbf{D}(\mathbf{r},t),$$

where the electric displacement field is defined as:

$$\mathbf{D} = \varepsilon_0 \mathbf{E} + \mathbf{P} = \varepsilon_0 \varepsilon_{\mathrm{r}} \mathbf{E},$$

where ε_0 is the electric constant, ε_{r} the relative static permittivity, and P is the polarization density. Substituting this form for D in the expression for displacement current, it has two components:

$$\mathbf{J}_{\mathrm{D}} = \varepsilon_0 \frac{\partial \mathbf{E}}{\partial t} + \frac{\partial \mathbf{P}}{\partial t}.$$

The first term on the right hand side is present everywhere, even in a vacuum. It doesn't involve any actual movement of charge, but it nevertheless has an associated magnetic field, as if it were an actual current. Some authors apply the name *displacement current* to only this contribution.

The second term on the right hand side is the displacement current as originally conceived by Maxwell, associated with the polarization of the individual molecules of the dielectric material.

Maxwell's original explanation for displacement current focused upon the situation that occurs in dielectric media. In the modern post-aether era, the concept has been extended to apply to situations with no material media present, for example, to the vacuum between the plates of a charging vacuum capacitor. The displacement current is justified today because it serves several requirements of an electromagnetic theory: correct prediction of magnetic fields in regions where no free current flows; prediction of wave propagation of electromagnetic fields; and conservation of electric charge in cases where charge density is time-varying.

Extending the Original Law: The Maxwell–Ampère Equation

Next, the circuital equation is extended by including the polarization current, thereby remedying the limited applicability of the original circuital law.

Treating free charges separately from bound charges, The equation including Maxwell's correction in terms of the H-field is (the H-field is used because it includes the magnetization currents, so J_M does not appear explicitly):

$$\oint_C \mathbf{H}\cdot \mathrm{d}\mathbf{l} = \iint_S \left(\mathbf{J}_{\mathrm{f}} + \frac{\partial \mathbf{D}}{\partial t} \right) \cdot \mathrm{d}\mathbf{S}$$

(integral form), where H is the magnetic H field (also called "auxiliary magnetic field", "magnetic field intensity", or just "magnetic field"), D is the electric displacement field, and J_f is the enclosed conduction current or free current density. In differential form,

$$\nabla \times \mathbf{H} = \mathbf{J}_{\mathrm{f}} + \frac{\partial \mathbf{D}}{\partial t}.$$

On the other hand, treating all charges on the same footing (disregarding whether they are bound or free charges), the generalized Ampère's equation, also called the Maxwell–Ampère equation, is in integral form:

$$\oint_C \mathbf{B}\cdot d\mathbf{l} = \iint_S \left(\mu_0 \mathbf{J} + \mu_0 \varepsilon_0 \frac{\partial \mathbf{E}}{\partial t} \right) \cdot d\mathbf{S}$$

In differential form,

$$\nabla \times \mathbf{B} = \mu_0 \mathbf{J} + \mu_0 \varepsilon_0 \frac{\partial \mathbf{E}}{\partial t}$$

In both forms J includes magnetization current density as well as conduction and polarization current densities. That is, the current density on the right side of the Ampère–Maxwell equation is:

$$\mathbf{J}_f + \mathbf{J}_D + \mathbf{J}_M = \mathbf{J}_f + \mathbf{J}_P + \mathbf{J}_M + \varepsilon_0 \frac{\partial \mathbf{E}}{\partial t} = \mathbf{J} + \varepsilon_0 \frac{\partial \mathbf{E}}{\partial t},$$

where current density J_D is the *displacement current*, and J is the current density contribution actually due to movement of charges, both free and bound. Because $\nabla \cdot D = \rho$, the charge continuity issue with Ampère's original formulation is no longer a problem. Because of the term in $\varepsilon_{0\partial}E/\partial t$, wave propagation in free space now is possible.

With the addition of the displacement current, Maxwell was able to hypothesize (correctly) that light was a form of electromagnetic wave.

Proof of Equivalence

Proof that the formulations of the circuital law in terms of free current are equivalent to the formulations involving total current.

In this proof, we will show that the equation

$$\nabla \times \mathbf{H} = \mathbf{J}_f + \frac{\partial \mathbf{D}}{\partial t}$$

is equivalent to the equation

$$\frac{1}{\mu_0}(\nabla \times \mathbf{B}) = \mathbf{J} + \varepsilon_0 \frac{\partial \mathbf{E}}{\partial t}.$$

Note that we are only dealing with the differential forms, not the integral forms, but that is sufficient since the differential and integral forms are equivalent in each case, by the Kelvin–Stokes theorem.

We introduce the polarization density P, which has the following relation to E and D:

$$\mathbf{D} = \varepsilon_0 \mathbf{E} + \mathbf{P}.$$

Next, we introduce the magnetization density M, which has the following relation to B and H:

$$\frac{1}{\mu_0}\mathbf{B} = \mathbf{H} + \mathbf{M}$$

and the following relation to the bound current:

$$\begin{aligned}\mathbf{J}_{\text{bound}} &= \nabla \times \mathbf{M} + \frac{\partial \mathbf{P}}{\partial t} \\ &= \mathbf{J}_{\text{M}} + \mathbf{J}_{\text{P}},\end{aligned}$$

where

$$\mathbf{J}_{\text{M}} = \nabla \times \mathbf{M},$$

is called the magnetization current density, and

$$\mathbf{J}_{\text{P}} = \frac{\partial \mathbf{P}}{\partial t},$$

is the polarization current density. Taking the equation for B:

$$\begin{aligned}\frac{1}{\mu_0}(\nabla \times \mathbf{B}) &= \nabla \times (\mathbf{H} + \mathbf{M}) \\ &= \nabla \times \mathbf{H} + \mathbf{J}_{\text{M}} \\ &= \mathbf{J}_{\text{f}} + \mathbf{J}_{\text{P}} + \varepsilon_0 \frac{\partial \mathbf{E}}{\partial t} + \mathbf{J}_{\text{M}}.\end{aligned}$$

Consequently, referring to the definition of the bound current:

$$\begin{aligned}\frac{1}{\mu_0}(\nabla \times \mathbf{B}) &= \mathbf{J}_{\text{f}} + \mathbf{J}_{\text{bound}} + \varepsilon_0 \frac{\partial \mathbf{E}}{\partial t} \\ &= \mathbf{J} + \varepsilon_0 \frac{\partial \mathbf{E}}{\partial t},\end{aligned}$$

as was to be shown.

Ampère's Circuital Law in cgs Units

In cgs units, the integral form of the equation, including Maxwell's correction, reads

$$\oint_C \mathbf{B} \cdot d\mathbf{l} = \frac{1}{c} \iint_S \left(4\pi \mathbf{J} + \frac{\partial \mathbf{E}}{\partial t} \right) \cdot d\mathbf{S},$$

where c is the speed of light.

The differential form of the equation (again, including Maxwell's correction) is,

$$\nabla \times \mathbf{B} = \frac{1}{c}\left(4\pi \mathbf{J} + \frac{\partial \mathbf{E}}{\partial t}\right).$$

Gauss's Law

In physics, Gauss's law, also known as Gauss's flux theorem, is a law relating the distribution of electric charge to the resulting electric field. The surface under consideration may be a closed one enclosing a volume such as a spherical surface.

The law was first formulated by Joseph-Louis Lagrange in 1773, followed by Carl Friedrich Gauss in 1813, both in the context of the attraction of ellipsoids. It is one of Maxwell's four equations, which form the basis of classical electrodynamics. Gauss's law can be used to derive Coulomb's law, and vice versa.

In words, Gauss's law states that: The net electric flux through any hypothetical closed surface is equal to $\frac{1}{\varepsilon_0}$ times the net electric charge within that closed surface.

Gauss's law has a close mathematical similarity with a number of laws in other areas of physics, such as Gauss's law for magnetism and Gauss's law for gravity. In fact, any inverse-square law can be formulated in a way similar to Gauss's law: for example, Gauss's law itself is essentially equivalent to the inverse-square Coulomb's law, and Gauss's law for gravity is essentially equivalent to the inverse-square Newton's law of gravity.

The law can be expressed mathematically using vector calculus in integral form and differential form; both are equivalent since they are related by the divergence theorem, also called Gauss's theorem. Each of these forms in turn can also be expressed two ways: In terms of a relation between the electric field E and the total electric charge, or in terms of the electric displacement field D and the *free* electric charge.

Equation involving the E Field

Gauss's law can be stated using either the electric field E or the electric displacement field D.

Integral Form

Gauss's law may be expressed as:

$$\Phi_E = \frac{Q}{\varepsilon_0}$$

where Φ_E is the electric flux through a closed surface S enclosing any volume V, Q is the total charge enclosed within V, and ε_0 is the electric constant. The electric flux Φ_E is defined as a surface integral of the electric field:

$$\Phi_E = \oiint_S \mathbf{E} \cdot d\mathbf{A}$$

where E is the electric field, dA is a vector representing an infinitesimal element of area of the surface, and represents the dot product of two vectors.

Since the flux is defined as an *integral* of the electric field, this expression of Gauss's law is called the *integral form.*

An important fact about this fundamental equation often doesn't find a mention in expositions that are not absolutely diligent. The above equation may fail to hold true in case the closed surface *S* contains a singularity of the electric field, which is physicists' term for a point in space where either a point charge exists and the field strength approaches infinity, or the field's magnitude or direction gets altered discontinuously due to the existence of a surface charge. In 2011, a modification of the above equation, called the Generalized Gauss's Theorem by its original creator, was published in the proceedings of the 2011 Annual Meeting of Electrostatics Society of America. The Generalized Gauss's Theorem allows the closed surface *S* to pass through singularities of the electric field. A corollary of the Generalized Gauss's Theorem, known as the simplest form of the Generalized Gauss's Theorem, holds true if the surface S is smooth. It states that,

$$\Phi_E = \frac{Q}{\varepsilon_0} + \frac{1}{2}\frac{Q'}{\varepsilon_0}$$

where *Q* is the net charge enclosed within *V* and *Q'* is the net charge contained by the closed surface *S* itself.

Applying the Integral Form

If the electric field is known everywhere, Gauss's law makes it possible to find the distribution of electric charge: The charge in any given region can be deduced by integrating the electric field to find the flux.

The reverse problem (when the electric charge distribution is known and the electric field must be computed) is much more difficult. The total flux through a given surface gives little information about the electric field, and can go in and out of the surface in arbitrarily complicated patterns.

An exception is if there is some symmetry in the problem, which mandates that the electric field passes through the surface in a uniform way. Then, if the total flux is known, the field itself can be deduced at every point. Common examples of symmetries which lend themselves to Gauss's law include: cylindrical symmetry, planar symmetry, and spherical symmetry.

Differential Form

By the divergence theorem, Gauss's law can alternatively be written in the *differential form*:

$$\nabla \cdot \mathbf{E} = \frac{\rho}{\varepsilon_0}$$

where $\nabla \cdot \mathbf{E}$ is the divergence of the electric field, ε_0 is the electric constant, and ρ is the total electric charge density (charge per unit volume).

Equivalence of Integral and Differential Forms

The integral and differential forms are mathematically equivalent, by the divergence theorem. Here is the argument more specifically.

Outline of Proof

The integral form of Gauss' law is:

$$\oiint_S = \mathbf{E} \cdot \mathrm{d}\mathbf{A} = \frac{Q}{\varepsilon_0}$$

for any closed surface S containing charge Q. By the divergence theorem, this equation is equivalent to:

$$\iiint_V \nabla \cdot \mathbf{E} \, \mathrm{d}V = \frac{Q}{\varepsilon_0}$$

for any volume V containing charge Q. By the relation between charge and charge density, this equation is equivalent to:

$$\iiint_V \nabla \cdot \mathbf{E} \, \mathrm{d}V = \iiint_V \frac{\rho}{\varepsilon_0} \mathrm{d}V$$

for any volume V. In order for this equation to be simultaneously true for every possible volume V, it is necessary (and sufficient) for the integrands to be equal everywhere. Therefore, this equation is equivalent to:

$$\nabla \cdot \mathbf{E} = \frac{\rho}{\varepsilon_0}.$$

Thus the integral and differential forms are equivalent.

Equation Involving the D Field

Free, Bound and Total Charge

The electric charge that arises in the simplest textbook situations would be classified as "free charge"—for example, the charge which is transferred in static electricity, or the charge on a capacitor plate. In contrast, "bound charge" arises only in the context of dielectric (polarizable) materials. (All materials are polarizable to some extent.) When such materials are placed in an external electric field, the electrons remain bound to their respective atoms, but shift a microscopic distance in response to the field, so that they're more on one side of the atom than the other. All these microscopic displacements add up to give a macroscopic net charge distribution, and this constitutes the "bound charge".

Although microscopically all charge is fundamentally the same, there are often practical reasons for wanting to treat bound charge differently from free charge. The result is that the more fundamental Gauss's law, in terms of E (above), is sometimes put into the equivalent form below, which is in terms of D and the free charge only.

Integral Form

This formulation of Gauss's law states the total charge form:

$$\Phi_D = Q_{\text{free}}$$

where Φ_D is the D-field flux through a surface S which encloses a volume V, and Q_{free} is the free charge contained in V. The flux Φ_D is defined analogously to the flux Φ_E of the electric field E through S:

$$\Phi_D = \oiint_S \mathbf{D} \cdot d\mathbf{A}$$

Differential Form

The differential form of Gauss's law, involving free charge only, states:

$$\nabla \cdot \mathbf{D} = \rho_{\text{free}}$$

where $\nabla \cdot$ D is the divergence of the electric displacement field, and ρ_{free} is the free electric charge density.

Equivalence of Total and Free Charge Statements

Proof that the formulations of Gauss's law in terms of free charge are equivalent to the formulations involving total charge.

In this proof, we will show that the equation

$$\nabla \cdot \mathbf{E} = \frac{\rho}{\varepsilon_0}$$

is equivalent to the equation

$$\nabla \cdot \mathbf{D} = \rho_{\text{free}}$$

Note that we are only dealing with the differential forms, not the integral forms, but that is sufficient since the differential and integral forms are equivalent in each case, by the divergence theorem.

We introduce the polarization density P, which has the following relation to E and D:

$$\mathbf{D} = \varepsilon_0 \mathbf{E} + \mathbf{P}$$

and the following relation to the bound charge:

$$\rho_{\text{bound}} = -\nabla \cdot \mathbf{P}$$

Now, consider the three equations:

$$\rho_{\text{bound}} = \nabla\cdot(-\mathbf{P})$$
$$\rho_{\text{free}} = \nabla\cdot\mathbf{D}$$
$$\rho = \nabla\cdot(\varepsilon_0\mathbf{E})$$

The key insight is that the sum of the first two equations is the third equation. This completes the proof: The first equation is true by definition, and therefore the second equation is true if and only if the third equation is true. So the second and third equations are equivalent, which is what we wanted to prove.

Equation for Linear Materials

In homogeneous, isotropic, nondispersive, linear materials, there is a simple relationship between E and D:

$$\mathbf{D} = \varepsilon\mathbf{E}$$

where ε is the permittivity of the material. For the case of vacuum (aka free space), $\varepsilon = \varepsilon_0$. Under these circumstances, Gauss's law modifies to

$$\Phi_E = \frac{Q_{\text{free}}}{\varepsilon}$$

for the integral form, and

$$\nabla\cdot\mathbf{E} = \frac{\rho_{\text{free}}}{\varepsilon}$$

for the differential form.

Interpretations

In Terms of Fields of Force

Gauss's theorem can be interpreted in terms of the lines of force of the field as follows:

The flux through a closed surface is dependent upon both the magnitude and direction of the electric field lines penetrating the surface. In general a positive flux is defined by these lines leaving the surface and negative flux by lines entering this surface. This results in positive charges causing a positive flux and negative charges creating a negative flux. These electric field lines will extend to infinite decreasing in strength by a factor of one over the distance from the source of the charge squared. The larger the number of field lines emanating from a charge the larger the magnitude of the charge is, and the closer together the field lines are the greater the magnitude of the electric filed. This has the natural result of the electric field becoming weaker as one moves away from a charged particle, but the surface area also increases so that the net electric field exiting this particle will stay the same. In other words the closed integral of the electric field and the dot product of the derivative of the area will equal the net charge enclosed divided by permittivity of free space.

Relation to Coulomb's Law

Deriving Gauss's Law from Coulomb's Law

Strictly speaking, Gauss's law cannot be derived from Coulomb's law alone, since Coulomb's law gives the electric field due to an individual point charge only. However, Gauss's law *can* be proven from Coulomb's law if it is assumed, in addition, that the electric field obeys the superposition principle. The superposition principle says that the resulting field is the vector sum of fields generated by each particle (or the integral, if the charges are distributed smoothly in space).

Outline of Proof

Coulomb's law states that the electric field due to a stationary point charge is:

$$\mathbf{E}(\mathbf{r}) = \frac{q}{4\pi\varepsilon_0}\frac{\mathbf{e}_r}{r^2}$$

where,

e_r is the radial unit vector,

r is the radius, |r|,

ε_0 is the electric constant,

q is the charge of the particle, which is assumed to be located at the origin.

Using the expression from Coulomb's law, we get the total field at r by using an integral to sum the field at r due to the infinitesimal charge at each other point s in space, to give,

$$\mathbf{E}(\mathbf{r}) = \frac{1}{4\pi\varepsilon_0}\int \frac{\rho(\mathbf{s})(\mathbf{r}-\mathbf{s})}{|\mathbf{r}-\mathbf{s}|^3}\,\mathrm{d}^3\mathbf{s}$$

where ρ is the charge density. If we take the divergence of both sides of this equation with respect to r, and use the known theorem,

$$\nabla\cdot\left(\frac{\mathbf{r}}{|\mathbf{r}|^3}\right) = 4\pi\delta(\mathbf{r})$$

where δ(r) is the Dirac delta function, the result is,

$$\nabla\cdot\mathbf{E}(\mathbf{r}) = \frac{1}{\varepsilon_0}\int \rho(\mathbf{s})\,\delta(\mathbf{r}-\mathbf{s})\,\mathrm{d}^3\mathbf{s}$$

Using the "sifting property" of the Dirac delta function, we arrive at,

$$\nabla\cdot\mathbf{E}(\mathbf{r}) = \frac{\rho(\mathbf{r})}{\varepsilon_0},$$

which is the differential form of Gauss' law, as desired.

Note that since Coulomb's law only applies to stationary charges, there is no reason to expect Gauss's law to hold for moving charges based on this derivation alone. In fact, Gauss's law does hold for moving charges, and in this respect Gauss's law is more general than Coulomb's law.

Deriving Coulomb's Law from Gauss's Law

Strictly speaking, Coulomb's law cannot be derived from Gauss's law alone, since Gauss's law does not give any information regarding the curl of E. However, Coulomb's law *can* be proven from Gauss's law if it is assumed, in addition, that the electric field from a point charge is spherically symmetric (this assumption, like Coulomb's law itself, is exactly true if the charge is stationary, and approximately true if the charge is in motion).

Outline of Proof

Taking S in the integral form of Gauss' law to be a spherical surface of radius r, centered at the point charge Q, we have,

$$\oint_S \mathrm{E} \cdot d\mathbf{A} = \frac{Q}{\varepsilon_0}$$

By the assumption of spherical symmetry, the integrand is a constant which can be taken out of the integral. The result is,

$$4\pi r^2 \hat{\mathbf{r}} \cdot \mathbf{E}(\mathbf{r}) = \frac{Q}{\varepsilon_0}$$

where $\hat{\mathrm{r}}$ is a unit vector pointing radially away from the charge. Again by spherical symmetry, E points in the radial direction, and so we get,

$$\mathbf{E}(\mathbf{r}) = \frac{Q}{4\pi\varepsilon_0} \frac{\hat{\mathbf{r}}}{r^2}$$

which is essentially equivalent to Coulomb's law. Thus the inverse-square law dependence of the electric field in Coulomb's law follows from Gauss' law.

Coulomb's Law

Coulomb's law, or Coulomb's inverse-square law, is an experimental law of physics that quantifies the amount of force between two stationary, electrically charged particles. The electric force between charged bodies at rest is conventionally called *electrostatic force* or Coulomb force. The quantity of electrostatic force between stationary charges is always described by Coulomb's law. The law was first published in 1785 by French physicist Charles-Augustin de Coulomb, and was essential to the development of the theory of electromagnetism, maybe even its starting point, because it was now possible to discuss quantity of electric charge in a meaningful way.

In its scalar form, the law is:

$$F = k_e \frac{q_1 q_2}{r^2},$$

where k_e is Coulomb's constant ($k_e \approx 9\times10^9$ N·m²·C⁻²), q_1 and q_2 are the signed magnitudes of the charges, and the scalar r is the distance between the charges. The force of the interaction between the charges is attractive if the charges have opposite signs (i.e., F is negative) and repulsive if like-signed (i.e., F is positive).

Being an inverse-square law, the law is analogous to Isaac Newton's inverse-square law of universal gravitation, but gravitational forces are always attractive, while electrostatic forces can be attractive or repulsive. Coulomb's law can be used to derive Gauss's law, and vice versa. The two laws are equivalent, expressing the same physical law in different ways. The law has been tested extensively, and observations have upheld the law on a scale from 10^{-16} m to 10^8 m.

Charles-Augustin de Coulomb.

Ancient cultures around the Mediterranean knew that certain objects, such as rods of amber, could be rubbed with cat's fur to attract light objects like feathers and papers. Thales of Miletus made a series of observations on static electricity around 600 BC, from which he believed that friction rendered amber magnetic, in contrast to minerals such as magnetite, which needed no rubbing. Thales was incorrect in believing the attraction was due to a magnetic effect, but later science would prove a link between magnetism and electricity. Electricity would remain little more than an intellectual curiosity for millennia until 1600, when the English scientist William Gilbert made a careful study of electricity and magnetism, distinguishing the lodestone effect from static electricity produced by rubbing amber.This association gave rise to the English words "electric" and "electricity", which made their first appearance in print in Thomas Browne's *Pseudodoxia Epidemica* of 1646.

Early investigators of the 18th century who suspected that the electrical force diminished with distance as the force of gravity did (i.e., as the inverse square of the distance) included Daniel Bernoulli and Alessandro Volta, both of whom measured the force between plates of a capacitor, and Franz Aepinus who supposed the inverse-square law in 1758.

Based on experiments with electrically charged spheres, Joseph Priestley of England was among the first to propose that electrical force followed an inverse-square law, similar to Newton's law of universal gravitation. However, he did not generalize or elaborate on this. In 1767, he conjectured that the force between charges varied as the inverse square of the distance.

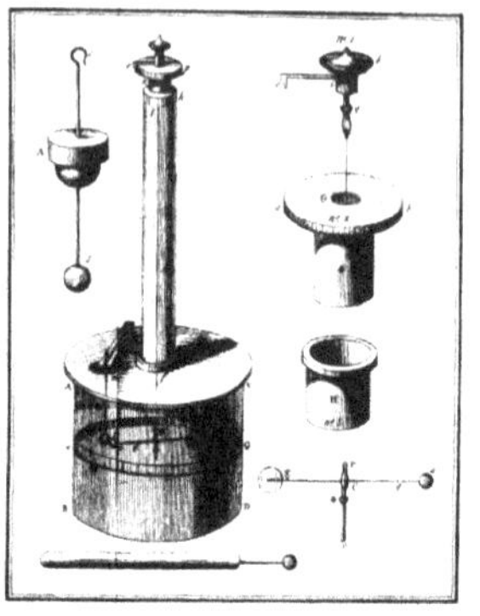

Coulomb's torsion balance.

In 1769, Scottish physicist John Robison announced that, according to his measurements, the force of repulsion between two spheres with charges of the same sign varied as $x^{-2.06}$.

In the early 1770s, the dependence of the force between charged bodies upon both distance and charge had already been discovered, but not published, by Henry Cavendish of England.

Finally, in 1785, the French physicist Charles-Augustin de Coulomb published his first three reports of electricity and magnetism where he stated his law. This publication was essential to the development of the theory of electromagnetism. He used a torsion balance to study the repulsion and attraction forces of charged particles, and determined that the magnitude of the electric force between two point charges is directly proportional to the product of the charges and inversely proportional to the square of the distance between them.

The torsion balance consists of a bar suspended from its middle by a thin fiber. The fiber acts as a very weak torsion spring. In Coulomb's experiment, the torsion balance was an insulating rod with a metal-coated ball attached to one end, suspended by a silk thread. The ball was charged with a known charge of static electricity, and a second charged ball of the same polarity was brought near it. The two charged balls repelled one another, twisting the fiber through a certain angle, which could be read from a scale on the instrument. By knowing how much force it took to twist the fiber through a given angle, Coulomb was able to calculate the force between the balls and derive his inverse-square proportionality law.

The Law

Coulomb's law states that:

> "The magnitude of the electrostatic force of attraction or repulsion between two point charges is directly proportional to the product of the magnitudes of charges and inversely proportional to the square of the distance between them".

The force is along the straight line joining them. If the two charges have the same sign, the electrostatic force between them is repulsive; if they have different signs, the force between them is attractive.

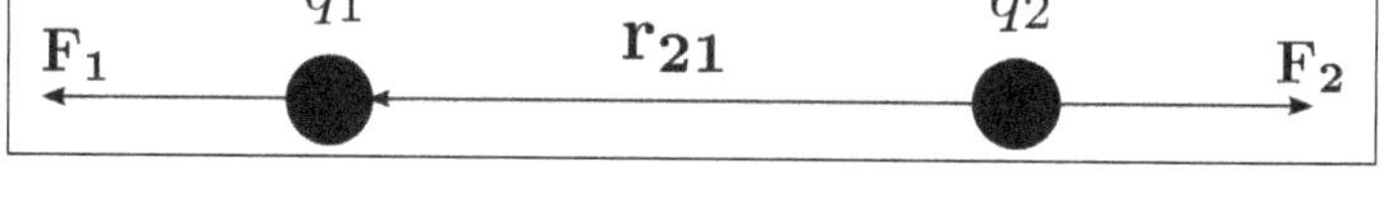

Coulomb's law can also be stated as a simple mathematical expression. The scalar and vector forms of the mathematical equation are

$$|\mathbf{F}| = k_e \frac{|q_1 q_2|}{r^2} \text{ and } \mathbf{F}_1 = k_e \frac{q_1 q_2}{|\mathbf{r}_{21}|^2} \hat{\mathbf{r}}_{21}, \text{ respectively,}$$

where k_e is Coulomb's constant ($k_e = 8.9875517873681764 \times 10^9$ N·m²·C⁻²), q_1 and q_2 are the signed magnitudes of the charges, the scalar r is the distance between the charges, the vector $r_{21} = r_1 - r_2$ is the vectorial distance between the charges, and $\hat{r}_{21} = \frac{r_{21}}{|r_{21}|}$ (a unit vector pointing from q_2 to q_1). The vector form of the equation calculates the force F_1 applied on q_1 by q_2. If r_{12} is used instead, then the effect on q_2 can be found. It can be also calculated using Newton's third law: $F_2 = -F_1$.

Units

When the electromagnetic theory is expressed in the International System of Units, force is measured in newtons, charge in coulombs, and distance in meters. Coulomb's constant is given by $k_e = 1/4\pi\varepsilon_0$. The constant ε_0 is the vacuum electric permittivity (also known as "electric constant") in $C^2 \cdot m^{-2} \cdot N^{-1}$. It should not be confused with ε_r, which is the dimensionless relative permittivity of the material in which the charges are immersed, or with their product $\varepsilon_a = \varepsilon_0 \varepsilon_r$, which is called "absolute permittivity of the material" and is still used in electrical engineering.

The SI derived units for the electric field are volts per meter, newtons per coulomb, or tesla meters per second.

Coulomb's law and Coulomb's constant can also be interpreted in various terms:

- Atomic units. In atomic units the force is expressed in hartrees per Bohr radius, the charge in terms of the elementary charge, and the distances in terms of the *Bohr radius*.
- Electrostatic units or Gaussian units. In electrostatic units and Gaussian units, the unit charge (*esu* or statcoulomb) is defined in such a way that the Coulomb constant k disappears because it has the value of one and becomes dimensionless.
- Lorentz–Heaviside units (also called *rationalized*). In Lorentz–Heaviside units the Coulomb constant is $k_e = 1/4\pi\varepsilon_0$ and becomes dimensionless.

Gaussian units and Lorentz–Heaviside units are both CGS unit systems. Gaussian units are more amenable for microscopic problems such as the electrodynamics of individual electrically charged particles. SI units are more convenient for practical, large-scale phenomena, such as engineering applications.

Electric Field

An electric field is a vector field that associates to each point in space the Coulomb force experienced by a test charge. In the simplest case, the field is considered to be generated solely by a single source point charge. The strength and direction of the Coulomb force F on a test charge q_t depends on the electric field E that it finds itself in, such that $F = q_t E$. If the field is generated by a positive source point charge q, the direction of the electric field points along lines directed radially outwards from it, i.e. in the direction that a positive point test charge q_t would move if placed in the field. For a negative point source charge, the direction is radially inwards.

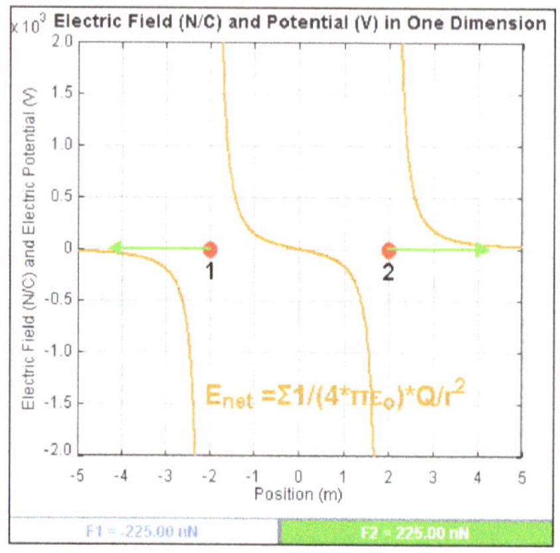

If two charges have the same sign, the electrostatic force between them is repulsive; if they have different sign, the force between them is attractive.

The magnitude of the electric field E can be derived from Coulomb's law. By choosing one of the point charges to be the source, and the other to be the test charge, it follows from Coulomb's law that the magnitude of the electric field E created by a single source point charge q at a certain distance from it r in vacuum is given by:

$$|\mathbf{E}| = \frac{1}{4\pi\varepsilon_0}\frac{|q|}{r^2}$$

Coulomb's Constant

Coulomb's constant is a proportionality factor that appears in Coulomb's law as well as in other electric-related formulas. The value of this constant is dependent upon the medium that the charged objects are immersed in. Denoted k_e, it is also called the electric force constant or electrostatic constant, hence the subscript e.

The exact value of Coulomb's constant in the case of air or vacuum is:

$$\begin{aligned} k_e &= \frac{1}{4\pi\varepsilon_0} = \frac{c_0^2\mu_0}{4\pi} = c_0^2 \times 10^{-7}\ \mathrm{H{\cdot}m^{-1}} \\ &= 8.987\,551\,787\,368\,176\,4 \times 10^9\ \mathrm{N{\cdot}m^2{\cdot}C^{-2}} \end{aligned}$$

Limitations

There are three conditions to be fulfilled for the validity of Coulomb's law:

- The charges must have a spherically symmetric distribution (e.g. be point charges, or a charged metal sphere).
- The charges must not overlap (e.g. they must be distinct point charges).
- The charges must be stationary with respect to each other.

The last of these is known as the electrostatic approximation. When movement takes place, Einstein's theory of relativity must be taken into consideration, and a result, an extra factor is introduced, which alters the force produced on the two objects. This extra part of the force is called the magnetic force, and is described by magnetic fields. For slow movement, the magnetic force is minimal and Coulomb's law can still be considered approximately correct, but when the charges are moving more quickly in relation to each other, the full electrodynamic rules (incorporating the magnetic force) must be considered.

Quantum Field Theory Origin

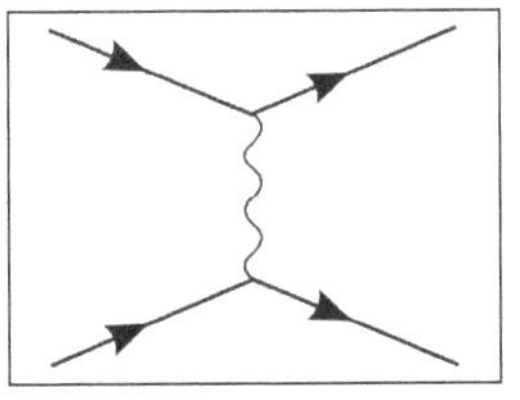

The most basic Feynman diagram for QED interaction between two fermions.

In simple terms, the Coulomb potential derives from the QED Lagrangian as follows. The Lagrangian of quantum electrodynamics is normally written in natural units, but in SI units, it is:

$$\mathcal{L}_{\text{QED}} = \bar{\psi}(i\hbar c\gamma^{\mu}D_{\mu} - mc^2)\psi - \frac{1}{4\hbar c}F_{\mu\nu}F^{\mu\nu}.$$

where the covariant derivative (in SI units) is:

$$D_{\mu} = \partial_{\mu} + \frac{ig}{\hbar c}A_{\mu}$$

where g is the gauge coupling parameter. By putting the covariant derivative into the lagrangian explicitly, the interaction term (the term involving both A and ψ) is seen to be:

$$\mathcal{L}_{\text{int}} = ig\bar{\psi}\gamma^{\mu}A_{\mu}\psi$$

The most basic Feynman diagram for a QED interaction between two fermions is the exchange of a single photon, with no loops. Following the Feynman rules, this therefore contributes two QED vertex factors ($igQ\gamma_{\mu}$) to the potential, where Q is the QED-charge operator (Q gives the charge in terms of the electron charge, and hence is exactly −1 for electrons, etc.). For the photon in the diagram, the Feynman rules demand the contribution of one bosonic massless propagator $\left(\frac{\hbar c}{k^2}\right)$. Ignoring the momentum on the external legs (the fermions), the potential is therefore:

$$V(\mathbf{r}) = \frac{1}{(2\pi)^3}\int e^{\frac{i\mathbf{k}\cdot\mathbf{r}}{\hbar}}(iQ_1g^2\gamma_{\mu})(iQ_2g^2\gamma_{\nu})\left(\frac{\hbar c}{k^2}\right)d^3k$$

which can be more usefully written as,

$$V(\mathbf{r}) = \frac{-g^2\hbar cQ_1Q_2}{4\pi}\frac{1}{(2\pi)^3}\int e^{\frac{i\mathbf{k}\cdot\mathbf{r}}{\hbar}}\frac{4\pi\eta_{\mu\nu}}{k^2}d^3k$$

where Q_i is the QED-charge on the ith particle. Recognising the integral as just being a Fourier transform enables the equation to be simplified:

$$V(\mathbf{r}) = \frac{-g^2 \hbar c Q_1 Q_2}{4\pi} \frac{1}{r}.$$

For various reasons, it is more convenient to define the fine-structure constant $\alpha = \frac{g^2}{4\pi}$, and then define $e = \sqrt{4\pi\alpha\varepsilon_0 \hbar c}$. Rearranging these definitions gives:

$$g^2 \hbar c = \frac{e^2}{\varepsilon_0}$$

In natural units ($\hbar = 1$, $c = 1$, and $\varepsilon_0 = 1$), $g = e$. Continuing in SI units, the potential is therefore

$$V(\mathrm{r}) = \frac{-e^2}{4\pi\varepsilon_0} \frac{Q_1 Q_2}{r}$$

Defining $q_i = eQ_i$, as the macroscopic 'electric charge', makes e the macroscopic 'electric charge' for an electron, and enables the formula to be put into the familiar form of the Coulomb potential:

$$V(\mathrm{r}) = \frac{-1}{4\pi\varepsilon_0} \frac{q_1 q_2}{r}$$

The force ($\frac{dV(\mathbf{r})}{dr}$) is therefore :

$$F(\mathbf{r}) = \frac{1}{4\pi\varepsilon_0} \frac{q_1 q_2}{r^2}$$

The derivation makes clear that the force law is only an approximation — it ignores the momentum of the input and output fermion lines, and ignores all quantum corrections (i.e. the myriad possible diagrams with internal loops).

The Coulomb potential, and its derivation, can be seen as a special case of the Yukawa potential (specifically, the case where the exchanged boson – the photon – has no rest mass).

Scalar Form

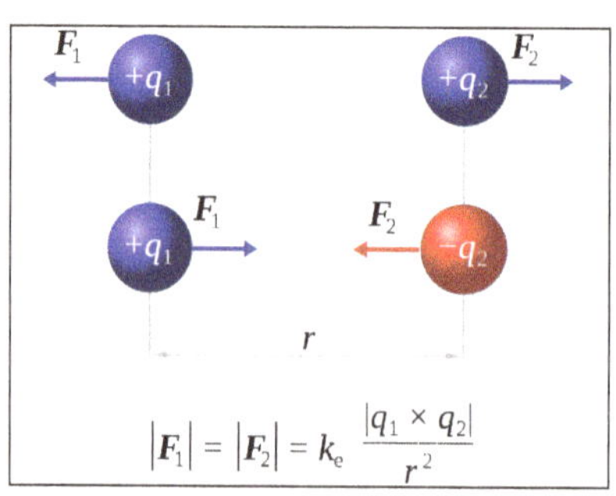

The absolute value of the force F between two point charges q and Q relates to the distance between the point charges and to the simple product of their charges. The diagram shows that like charges repel each other, and opposite charges mutually attract.

When it is of interest to know the magnitude of the electrostatic force (and not its direction), it may be easiest to consider a scalar version of the law. The scalar form of Coulomb's Law relates the magnitude and sign of the electrostatic force F acting simultaneously on two point charges q_1 and q_2 as follows:

$$|\mathbf{F}| = k_e \frac{|q_1 q_2|}{r^2}$$

where r is the separation distance and k_e is Coulomb's constant. If the product q_1q_2 is positive, the force between the two charges is repulsive; if the product is negative, the force between them is attractive.

Vector Form

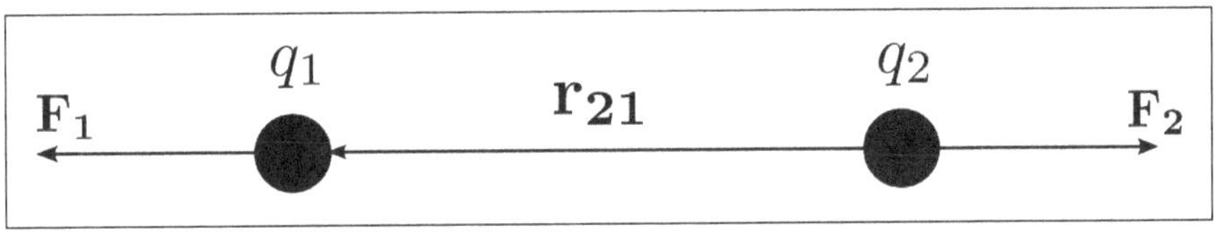

In the image, the vector F_1 is the force experienced by q_1, and the vector F_2 is the force experienced by q_2. When $q_1q_2 > 0$ the forces are repulsive (as in the image) and when $q_1q_2 < 0$ the forces are attractive (opposite to the image). The magnitude of the forces will always be equal.

Coulomb's law states that the electrostatic force F_1 experienced by a charge, q_1 at position r_1, in the vicinity of another charge, q_2 at position r_2, in a vacuum is equal to:

$$\mathbf{F}_1 = \frac{q_1 q_2}{4\pi\varepsilon_0} \frac{(\mathbf{r}_1 - \mathbf{r}_2)}{|\mathbf{r}_1 - \mathbf{r}_2|^3} = \frac{q_1 q_2}{4\pi\varepsilon_0} \frac{\hat{\mathbf{r}}_{21}}{|\mathbf{r}_{21}|^2},$$

where $r_{21} = r_1 - r_2$, the unit vector $\hat{r}_{21} = \frac{r_{21}}{|r_{21}|}$, and ε_0 is the electric constant.

The vector form of Coulomb's law is simply the scalar definition of the law with the direction given by the unit vector, $\hat{r}_{21}$, parallel with the line *from* charge q_2 *to* charge q_1. If both charges have the same sign (like charges) then the product q_1q_2 is positive and the direction of the force on q_1 is given by $\hat{r}_{21}$; the charges repel each other. If the charges have opposite signs then the product q_1q_2 is negative and the direction of the force on q_1 is given by $-\hat{r}_{21} = \hat{r}_{12}$; the charges attract each other.

The electrostatic force F_2 experienced by q_2, according to Newton's third law, is $F_2 = -F_1$.

System of Discrete Charges

The law of superposition allows Coulomb's law to be extended to include any number of point charges. The force acting on a point charge due to a system of point charges is simply the vector addition of the individual forces acting alone on that point charge due to each one of the charges. The resulting force vector is parallel to the electric field vector at that point, with that point charge removed.

The force F on a small charge q at position r, due to a system of N discrete charges in vacuum is:

$$\mathbf{F}(\mathbf{r}) = \frac{q}{4\pi\varepsilon_0}\sum_{i=1}^{N} q_i \frac{\mathbf{r}-\mathbf{r_i}}{|\mathbf{r}-\mathbf{r_i}|^3} = \frac{q}{4\pi\varepsilon_0}\sum_{i=1}^{N} q_i \frac{\widehat{\mathbf{R_i}}}{|\mathbf{R_i}|^2},$$

where q_i and r_i are the magnitude and position respectively of the ith charge, $\hat{R}_i$ is a unit vector in the direction of $R_i = r - r_i$ (a vector pointing from charges q_i to q).

Continuous Charge Distribution

In this case, the principle of linear superposition is also used. For a continuous charge distribution, an integral over the region containing the charge is equivalent to an infinite summation, treating each infinitesimal element of space as a point charge dq. The distribution of charge is usually linear, surface or volumetric.

For a linear charge distribution (a good approximation for charge in a wire) where $\lambda(r')$ gives the charge per unit length at position r′, and $d\ell'$ is an infinitesimal element of length,

$$dq = \lambda(\mathbf{r}')\,d\ell'.$$

For a surface charge distribution (a good approximation for charge on a plate in a parallel plate capacitor) where $\sigma(r')$ gives the charge per unit area at position r′, and dA' is an infinitesimal element of area,

$$dq = \sigma(\mathbf{r}')\,dA'.$$

For a volume charge distribution (such as charge within a bulk metal) where $\rho(r')$ gives the charge per unit volume at position r′, and dV' is an infinitesimal element of volume,

$$dq = \rho(\mathbf{r}')\,dV'.$$

The force on a small test charge q' at position r in vacuum is given by the integral over the distribution of charge:

$$\mathbf{F} = \frac{q'}{4\pi\varepsilon_0}\int dq \frac{\mathbf{r}-\mathbf{r}'}{|\mathbf{r}-\mathbf{r}'|^3}.$$

Simple Experiment to Verify Coulomb's Law

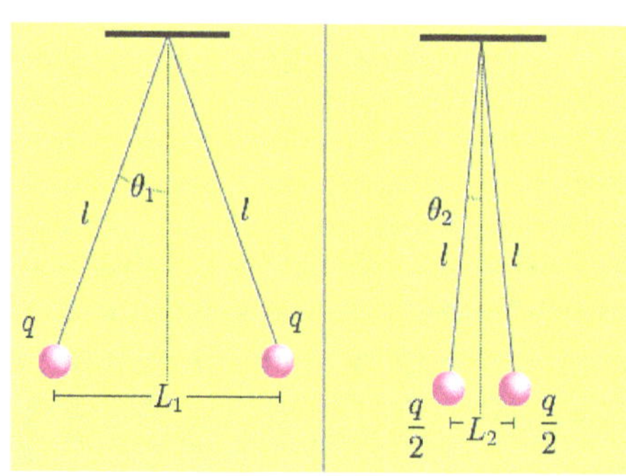

Experiment to verify Coulomb's law.

It is possible to verify Coulomb's law with a simple experiment. Consider two small spheres of mass m and same-sign charge q, hanging from two ropes of negligible mass of length l. The forces acting on each sphere are three: the weight mg, the rope tension T and the electric force F.

In the equilibrium state:

$$T \sin\theta_1 = F_1$$

and:

$$T \cos\theta_1 = mg$$

Dividing $T \sin\theta_1 = F_1$ by, $T \cos\theta_1 = mg$:

$$\frac{\sin\theta_1}{\cos\theta_1} = \frac{F_1}{mg} \Rightarrow F_1 = mg \tan\theta_1$$

Let L_1 be the distance between the charged spheres; the repulsion force between them F_1, assuming Coulomb's law is correct, is equal to:

$$F_1 = \frac{q^2}{4\pi\varepsilon_0 L_1^2}$$

so:

$$\frac{q^2}{4\pi\varepsilon_0 L_1^2} = mg \tan\theta_1$$

If we now discharge one of the spheres, and we put it in contact with the charged sphere, each one of them acquires a charge $q/2$. In the equilibrium state, the distance between the charges will be $L_2 < L_1$ and the repulsion force between them will be:

$$F_2 = \frac{(\frac{q}{2})^2}{4\pi\varepsilon_0 L_2^2} = \frac{\frac{q^2}{4}}{4\pi\varepsilon_0 L_2^2}$$

We know that $F_2 = mg \tan\theta_2$

and

$$\frac{\frac{q^2}{4}}{4\pi\varepsilon_0 L_2^2} = mg \tan\theta_2$$

Dividing $\frac{\quad}{\pi\varepsilon_0 \quad 1}$ $mg \tan$ by $F_2 = \frac{(\frac{q}{2})^2}{4\pi\varepsilon_0 L_2^2} = \frac{\frac{q^2}{4}}{4\pi\varepsilon_0 L_2^2}$, we get: $$\frac{\left(\dfrac{q^2}{4\pi\varepsilon_0 L_1^2}\right)}{\left(\dfrac{\dfrac{q^2}{4}}{4\pi\varepsilon_0 L_2^2}\right)} = \frac{mg\tan\theta_1}{mg\tan\theta_2} \Rightarrow 4\left(\frac{L_2}{L_1}\right)^2 = \frac{\tan\theta_1}{\tan\theta_2}$$

Measuring the angles θ_1 and θ_2 and the distance between the charges L_1 and L_2 is sufficient to verify that the equality is true taking into account the experimental error. In practice, angles can be difficult to measure, so if the length of the ropes is sufficiently great, the angles will be small enough to make the following approximation:

$$\tan\theta \approx \sin\theta = \frac{\frac{L}{2}}{\ell} = \frac{L}{2\ell} \Rightarrow \frac{\tan\theta_1}{\tan\theta_2} \approx \frac{\frac{L_1}{2\ell}}{\frac{L_2}{2\ell}}$$

Using this approximation, the relationship (6) becomes the much simpler expression:

$$\frac{\frac{L_1}{2\ell}}{\frac{L_2}{2\ell}} \approx 4\left(\frac{L_2}{L_1}\right)^2 \Rightarrow \frac{L_1}{L_2} \approx 4\left(\frac{L_2}{L_1}\right)^2 \Rightarrow \frac{L_1}{L_2} \approx \sqrt[3]{4}$$

In this way, the verification is limited to measuring the distance between the charges and check that the division approximates the theoretical value.

Atomic Forces

Coulomb's law holds even within atoms, correctly describing the force between the positively charged atomic nucleus and each of the negatively charged electrons. This simple law also correctly accounts for the forces that bind atoms together to form molecules and for the forces that bind atoms and molecules together to form solids and liquids. Generally, as the distance between ions increases, the force of attraction, and binding energy, approach zero and ionic bonding is less favorable. As the magnitude of opposing charges increases, energy increases and ionic bonding is more favorable.

Faraday's Law of Electromagnetic Induction

Faraday's law of induction (briefly, Faraday's law) is a basic law of electromagnetism predicting how a magnetic field will interact with an electric circuit to produce an electromotive force (EMF)—a phenomenon called electromagnetic induction. It is the fundamental operating principle of transformers, inductors, and many types of electrical motors, generators and solenoids.

The Maxwell–Faraday equation (listed as one of Maxwell's equations) describes the fact that a spatially varying (and also possibly time-varying, depending on how a magnetic field varies in time) electric field always accompanies a time-varying magnetic field, while Faraday's law states that there is EMF (electromotive force, defined as electromagnetic work done on a unit charge when it has traveled one round of a conductive loop) on the conductive loop when the magnetic flux through the surface enclosed by the loop varies in time.

Faraday's law had been discovered and one aspect of it (transformer EMF) was formulated as the Maxwell–Faraday equation later. The equation of Faraday's law can be derived by the Maxwell–Faraday equation (describing transformer EMF) and the Lorentz force (describing motional EMF). The integral form of the Maxwell–Faraday equation describes only the transformer EMF, while the equation of Faraday's law describes both the transformer EMF and the motional EMF.

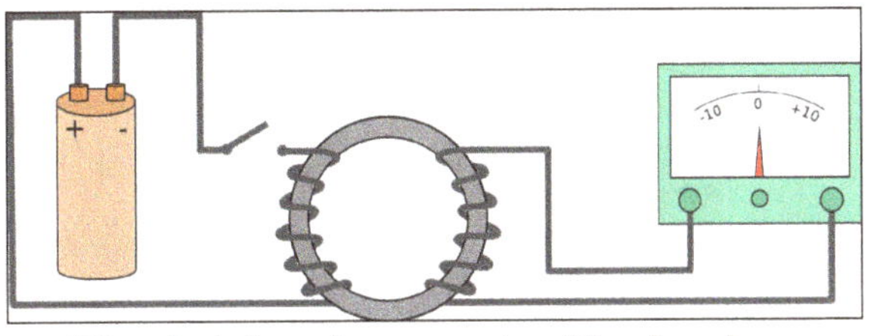

A diagram of Faraday's iron ring apparatus. The changing magnetic flux of the left coil induces a current in the right coil.

Faraday's disk, the first electric generator, a type of homopolar generator.

Electromagnetic induction was discovered independently by Michael Faraday in 1831 and Joseph Henry in 1832. Faraday was the first to publish the results of his experiments. In Faraday's first experimental demonstration of electromagnetic induction, he wrapped two wires around opposite sides of an iron ring (torus) (an arrangement similar to a modern toroidal transformer). Based on his assessment of recently discovered properties of electromagnets, he expected that when current started to flow in one wire, a sort of wave would travel through the ring and cause some electrical effect on the opposite side. He plugged one wire into a galvanometer, and watched it as he connected the other wire to a battery. Indeed, he saw a transient current (which he called a "wave of electricity") when he connected the wire to the battery, and another when he disconnected it. This induction was due to the change in magnetic flux that occurred when the battery was connected and disconnected. Within two months, Faraday had found several other manifestations of electromagnetic induction. For example, he saw transient currents when he quickly slid a bar magnet in and out of a coil of wires, and he generated a steady (DC) current by rotating a copper disk near the bar magnet with a sliding electrical lead ("Faraday's disk").

Michael Faraday explained electromagnetic induction using a concept he called lines of force. However, scientists at the time widely rejected his theoretical ideas, mainly because they were not formulated mathematically. An exception was James Clerk Maxwell, who in 1861–62 used Faraday's ideas as the basis of his quantitative electromagnetic theory. In Maxwell's papers, the time-varying aspect of electromagnetic induction is expressed as a differential equation which Oliver Heaviside referred to as Faraday's law even though it is different from the original version of Faraday's law, and does not describe motional EMF. Heaviside's version is the form recognized today in the group of equations known as Maxwell's equations.

Lenz's law, formulated by Emil Lenz in 1834, describes "flux through the circuit", and gives the direction of the induced EMF and current resulting from electromagnetic induction.

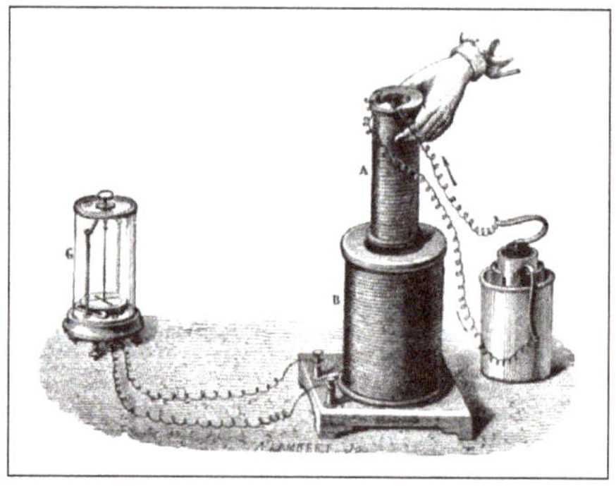

Faraday's experiment showing induction between coils of wire: The liquid battery *(right)* provides a current which flows through the small coil *(A)*, creating a magnetic field. When the coils are stationary, no current is induced. But when the small coil is moved in or out of the large coil *(B)*, the magnetic flux through the large coil changes, inducing a current which is detected by the galvanometer *(G)*.

Faraday's Law

The most widespread version of Faraday's law states:

> The electromotive force around a closed path is equal to the negative of the time rate of change of the magnetic flux enclosed by the path.

The closed path here is, in fact, conductive.

Mathematical Statement

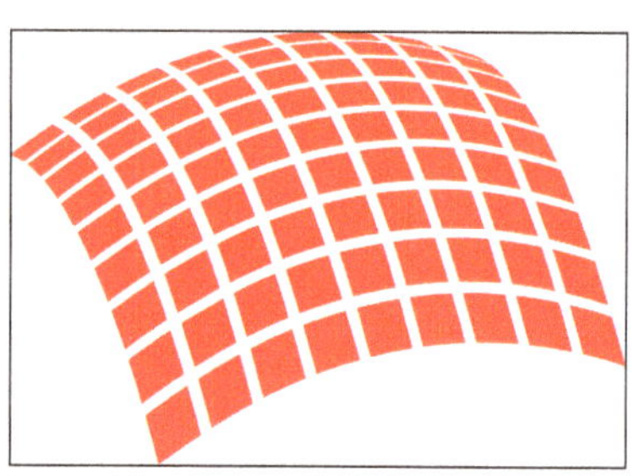

The definition of surface integral relies on splitting the surface Σ into small surface elements. Each element is associated with a vector dA of magnitude equal to the area of the element and with direction normal to the element and pointing "outward" (with respect to the orientation of the surface).

For a loop of wire in a magnetic field, the magnetic flux Φ_B is defined for any surface Σ whose boundary is the given loop. Since the wire loop may be moving, we write $\Sigma(t)$ for the surface. The magnetic flux is the surface integral:

$$\Phi_B = \iint_{\Sigma(t)} \mathbf{B}(t) \cdot \mathrm{d}\mathbf{A},$$

where dA is an element of surface area of the moving surface $\Sigma(t)$, B is the magnetic field, and B·dA is a vector dot product representing the element of flux through dA. In more visual terms, the magnetic flux through the wire loop is proportional to the number of magnetic flux lines that pass through the loop.

When the flux changes—because B changes, or because the wire loop is moved or deformed, or both—Faraday's law of induction says that the wire loop acquires an EMF, E, defined as the energy available from a unit charge that has travelled once around the wire loop. (Note that different textbooks may give different definitions. The set of equations used throughout the text was chosen to be compatible with the special relativity theory.) Equivalently, it is the voltage that would be measured by cutting the wire to create an open circuit, and attaching a voltmeter to the leads.

Faraday's law states that the EMF is also given by the rate of change of the magnetic flux:

$$\mathcal{E} = -\frac{\mathrm{d}\Phi_B}{\mathrm{d}t},$$

where $\mathcal{E}$ is the electromotive force (EMF) and Φ_B is the magnetic flux.

The direction of the electromotive force is given by Lenz's law.

The laws of induction of electric currents in mathematical form was established by Franz Ernst Neumann in 1845.

Faraday's law contains the information about the relationships between both the magnitudes and the directions of its variables. However, the relationships between the directions are not explicit; they are hidden in the mathematical formula.

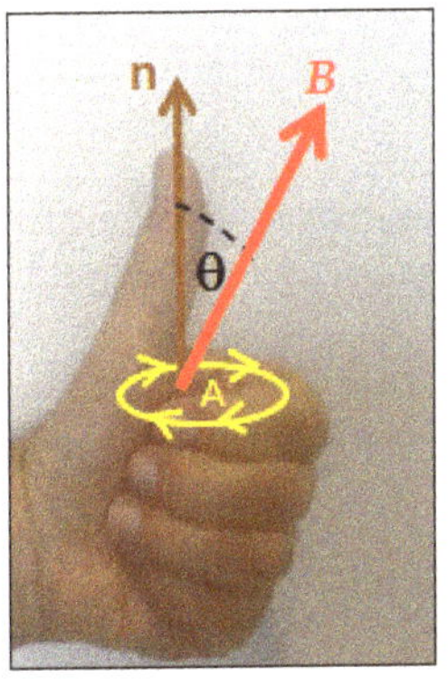

A Left Hand Rule for Faraday's Law. The sign of $\Delta\Phi_B$, the change in flux, is found based on the relationship between the magnetic field B, the area of the loop A, and the normal n to that area, as represented by the fingers of the left hand. If $\Delta\Phi_B$ is positive, the direction of the EMF is the same as that of the curved fingers (yellow arrowheads). If $\Delta\Phi_B$ is negative, the direction of the EMF is against the arrowheads.

It is possible to find out the direction of the electromotive force (EMF) directly from Faraday's law, without invoking Lenz's law. A left hand rule helps doing that, as follows:

- Align the curved fingers of the left hand with the loop (yellow line).
- Stretch your thumb. The stretched thumb indicates the direction of n (brown), the normal to the area enclosed by the loop.
- Find the sign of $\Delta\Phi_B$, the change in flux. Determine the initial and final fluxes (whose difference is $\Delta\Phi_B$) with respect to the normal n, as indicated by the stretched thumb.
- If the change in flux, $\Delta\Phi_B$, is positive, the curved fingers show the direction of the electromotive force (yellow arrowheads).
- If $\Delta\Phi_B$ is negative, the direction of the electromotive force is opposite to the direction of the curved fingers (opposite to the yellow arrowheads).

For a tightly wound coil of wire, composed of N identical turns, each with the same Φ_B, Faraday's law of induction states that,

$$\mathcal{E} = -N\frac{d\Phi_B}{dt}$$

where N is the number of turns of wire and Φ_B is the magnetic flux through a single loop.

Maxwell–Faraday equation

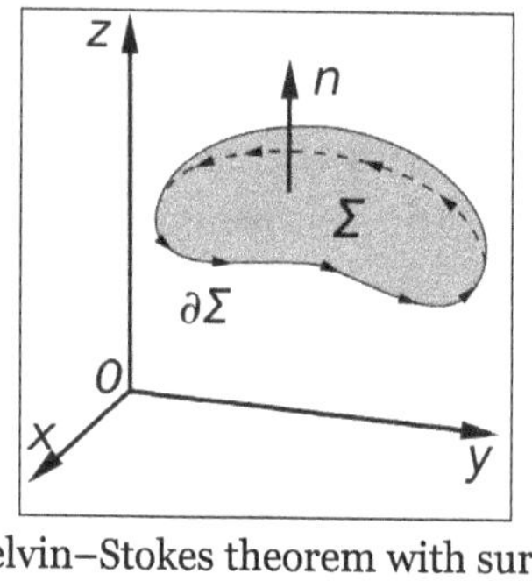

An illustration of the Kelvin–Stokes theorem with surface Σ, its boundary ∂Σ, and orientation n set by the right-hand rule.

The Maxwell–Faraday equation states that a time-varying magnetic field always accompanies a spatially varying (also possibly time-varying), non-conservative electric field, and vice versa. The Maxwell–Faraday equation is,

$$\nabla \times \mathbf{E} = -\frac{\partial \mathbf{B}}{\partial t}$$

(in SI units) where $\nabla \times$ is the curl operator and again E(r, t) is the electric field and B(r, t) is the magnetic field. These fields can generally be functions of position r and time t.

The Maxwell–Faraday equation is one of the four Maxwell's equations, and therefore plays a fundamental role in the theory of classical electromagnetism. It can also be written in an integral form by the Kelvin–Stokes theorem, thereby reproducing Faraday's law:

$$\oint_{\partial\Sigma} \mathbf{E}\cdot d\mathbf{l} = -\int_{\Sigma} \frac{\partial \mathbf{B}}{\partial t}\cdot d\mathbf{A}$$

where, as indicated in the figure:

- Σ is a surface bounded by the closed contour ∂Σ,
- E is the electric field, B is the magnetic field,
- dl is an infinitesimal vector element of the contour ∂Σ,
- dA is an infinitesimal vector element of surface Σ. If its direction is orthogonal to that surface patch, the magnitude is the area of an infinitesimal patch of surface.

For a planar surface Σ, a positive path element dl of curve ∂Σ is defined by the right-hand rule as one that points with the fingers of the right hand when the thumb points in the direction of the normal n to the surface Σ.

The integral around ∂Σ is called a path integral or line integral.

Notice that a nonzero path integral for E is different from the behavior of the electric field generated by charges. A charge-generated E-field can be expressed as the gradient of a scalar field that is a solution to Poisson's equation, and has a zero path integral.

The integral equation is true for *any* path ∂Σ through space, and any surface Σ for which that path is a boundary.

If the surface Σ is not changing in time, the equation can be rewritten:

$$\oint_{\partial\Sigma} \mathbf{E}\cdot d\mathbf{l} = -\frac{d}{dt}\int_{\Sigma} \mathbf{B}\cdot d\mathbf{A}.$$

The surface integral at the right-hand side is the explicit expression for the magnetic flux Φ_B through Σ.

The electric vector field induced by a changing magnetic flux, the solenoidal component of the overall electric field, can be approximated in the non-relativistic limit by the following volume integral equation:

$$\mathbf{E}_s(\mathbf{r}) \approx -\frac{1}{4\pi}\iiint_V \frac{(\frac{\partial \mathbf{B}}{\partial t}dV)\times\hat{\mathbf{r}}'}{|\mathbf{r}'|^2}$$

Proof

The four Maxwell's equations (including the Maxwell–Faraday equation), along with Lorentz force law, are a sufficient foundation to derive *everything* in classical electromagnetism. Therefore, it is possible to "prove" Faraday's law starting with these equations.

The starting point is the time-derivative of flux through an arbitrary surface Σ (that can move or be deformed) in space:

$$\frac{\mathrm{d}\Phi_B}{\mathrm{d}t} = \frac{\mathrm{d}}{\mathrm{d}t}\int_{\Sigma(t)} \mathbf{B}(t)\cdot \mathrm{d}\mathbf{A}$$

(by definition). This total time derivative can be evaluated and simplified with the help of the Maxwell–Faraday equation and some vector identities; the details are in the box below:

Consider the time-derivative of magnetic flux through a closed boundary (loop) that can move or be deformed. The area bounded by the loop is denoted as Σ(t)), then the time-derivative can be expressed as

$$\frac{\mathrm{d}\Phi_B}{\mathrm{d}t} = \frac{\mathrm{d}}{\mathrm{d}t}\int_{\Sigma(t)} \mathbf{B}(t)\cdot \mathrm{d}\mathbf{A}$$

The integral can change over time for two reasons: The integrand can change, or the integration region can change. These add linearly, therefore:

$$\left.\frac{\mathrm{d}\Phi_B}{\mathrm{d}t}\right|_{t=t_0} = \left(\int_{\Sigma(t_0)} \left.\frac{\partial \mathbf{B}}{\partial t}\right|_{t=t_0} \cdot \mathrm{d}\mathbf{A}\right) + \left(\frac{\mathrm{d}}{\mathrm{d}t}\int_{\Sigma(t)} \mathbf{B}(t_0)\cdot \mathrm{d}\mathbf{A}\right)$$

where t_0 is any given fixed time. We will show that the first term on the right-hand side corresponds to transformer EMF, the second to motional EMF (from the magnetic Lorentz force on charge carriers due to the motion or deformation of the conducting loop in the magnetic field). The first term on the right-hand side can be rewritten using the integral form of the Maxwell–Faraday equation:

$$\int_{\Sigma(t_0)} \left.\frac{\partial \mathbf{B}}{\partial t}\right|_{t=t_0} \cdot \mathrm{d}\mathbf{A} = -\oint_{\partial\Sigma(t_0)} \mathbf{E}(t_0)\cdot \mathrm{d}\mathbf{l}$$

Next, we analyze the second term on the right-hand side:

$$\frac{\mathrm{d}}{\mathrm{d}t}\int_{\Sigma(t)} \mathbf{B}(t_0)\cdot \mathrm{d}\mathbf{A}$$

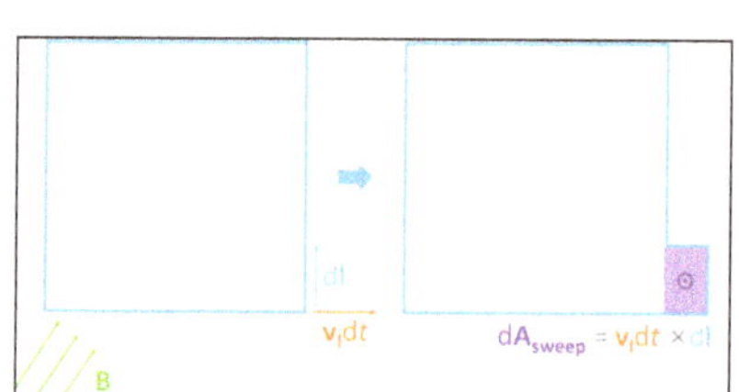

The area swept out by a vector element dl of a loop ∂Σ in time dt when it has moved with velocity v_l.

The proof of this is a little more difficult than the first term; more details and alternate approaches for the proof can be found in the references. As the loop moves and/or deforms, it sweeps out a surface. As a small part of the loop dl moves with velocity v_l over a short time dt, it sweeps out an area which vector is $dA_{sweep} = v_l\ dt \times dl$. Therefore, the change of the magnetic flux through the loop due to the deformation or movement of the loop over the time dt is $\mathbf{d}\Phi_B = \int \mathbf{B}\cdot\mathbf{dA}_{sweep} = \int \mathbf{B}\cdot(\mathbf{v}_l dt \times \mathrm{d}\mathbf{l}) = -\int dt\mathrm{d}\mathbf{l}\cdot(\mathbf{v}_l \times \mathbf{B})$.

Here, identities of triple scalar products are used. Therefore,

$$\frac{\mathrm{d}}{\mathrm{d}t}\int_{\Sigma(t)} \mathbf{B}(t_0)\cdot\mathrm{d}\mathbf{A} = -\oint_{\partial\Sigma(t_0)} (\mathbf{v}_l(t_0)\times\mathbf{B}(t_0))\cdot\mathrm{d}\mathbf{l}$$

where v_l is the velocity of a part of the loop $\partial\Sigma$.

Putting these together results in,

$$\left.\frac{\mathrm{d}\Phi_B}{\mathrm{d}t}\right|_{t=t_0} = \left(-\oint_{\partial\Sigma(t_0)} \mathbf{E}(t_0)\cdot\mathrm{d}\mathbf{l}\right) + \left(-\oint_{\partial\Sigma(t_0)} \big(\mathbf{v}_l(t_0)\times\mathbf{B}(t_0)\big)\cdot\mathrm{d}\mathbf{l}\right)$$

$$\left.\frac{\mathrm{d}\Phi_B}{\mathrm{d}t}\right|_{t=t_0} = -\oint_{\partial\Sigma(t_0)} \big(\mathbf{E}(t_0) + \mathbf{v}_l(t_0)\times\mathbf{B}(t_0)\big)\cdot\mathrm{d}\mathbf{l}.$$

The result is:

$$\frac{\mathrm{d}\Phi_B}{\mathrm{d}t} = -\oint_{\partial\Sigma}\left(\mathbf{E} + \mathbf{v}_l \times \mathbf{B}\right)\cdot\mathrm{d}\mathbf{l}.$$

where $\partial\Sigma$ is the boundary (loop) of the surface Σ, and v_l is the velocity of a part of the boundary.

In the case of a conductive loop, EMF (Electromotive Force) is the electromagnetic work done on a unit charge when it has traveled around the loop once, and this work is done by the Lorentz force. Therefore, EMF is expressed as,

$$\mathcal{E} = \oint\left(\mathbf{E} + \mathbf{v}\times\mathbf{B}\right)\cdot\mathrm{d}\mathbf{l}$$

where $\mathcal{E}$ is EMF and v is the unit charge velocity.

In a macroscopic view, for charges on a segment of the loop, v consists of two components in average; one is the velocity of the charge along the segment v_t, and the other is the velocity of the segment v_l (the loop is deformed or moved). v_t does not contribute to the work done on the charge since the direction of v_t is same to the direction of $d\mathbf{l}$. Mathematically,

$$(\mathbf{v}\times B)\cdot d\mathbf{l} = ((\mathbf{v}_t + \mathbf{v}_l)\times B)\cdot d\mathbf{l} = (\mathbf{v}_t \times B + \mathbf{v}_l \times B)\cdot d\mathbf{l} = (\mathbf{v}_l \times B)\cdot d\mathbf{l}$$

since $(\mathbf{v}_t \times B)$ is perpendicular to $d\mathbf{l}$ as $\mathbf{v}_t$ and $d\mathbf{l}$ are along the same direction. Now we can see that, for the conductive loop, EMF is same to the time-derivative of the magnetic flux through the loop except for the sign on it. Therefore, we now reach the equation of Faraday's law (for the conductive loop) as,

$$\frac{d\Phi_B}{dt} = -\mathcal{E}$$

where $\mathcal{E} = \oint (\mathbf{E} + \mathbf{v}_l \times \mathbf{B}) \cdot d\mathbf{l}$. With breaking this integral, $\oint \mathbf{E} \cdot d\mathbf{l}$ is for the transformer EMF (due to a time-varying magnetic field) and $\oint (\mathbf{v}_l \times \mathbf{B}) \cdot d\mathbf{l}$ is for the motional EMF (due to the magnetic Lorentz force on charges by the motion or deformation of the loop in the magnetic field).

EMF for Non-thin-wire Circuits

It is tempting to generalize Faraday's law to state: *If ∂Σ is any arbitrary closed loop in space whatsoever, then the total time derivative of magnetic flux through Σ equals the EMF around ∂Σ.* This statement, however, is not always true and the reason is not just from the obvious reason that EMF is undefined in empty space when no conductor is present., Faraday's law is not guaranteed to work unless the velocity of the abstract curve ∂Σ matches the actual velocity of the material conducting the electricity. The two examples illustrated below show that one often obtains incorrect results when the motion of ∂Σ is divorced from the motion of the material.

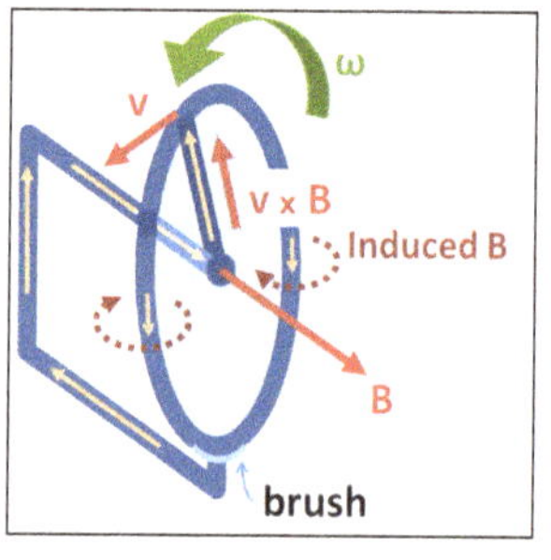

Faraday's homopolar generator. The disc rotates with angular rate ω, sweeping the conducting radius circularly in the static magnetic field B (which direction is along the disk surface normal). The magnetic Lorentz force v × B drives a current along the conducting radius to the conducting rim, and from there the circuit completes through the lower brush and the axle supporting the disc. This device generates an EMF and a current, although the shape of the "circuit" is constant and thus the flux through the circuit does not change with time.

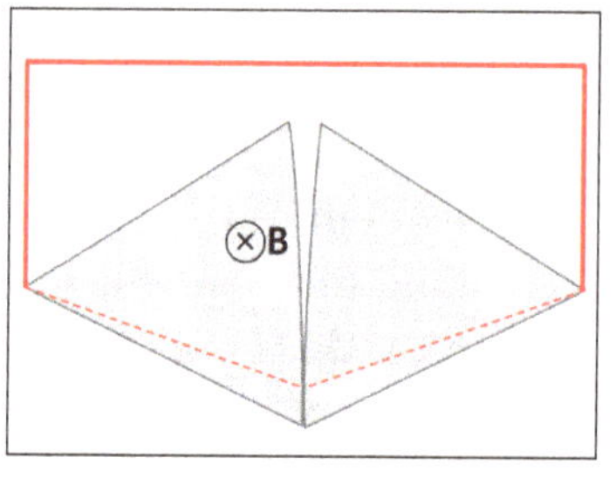

A wire (solid red lines) connects to two touching metal plates (silver) to form a circuit. The whole system sits in a uniform magnetic field, normal to the page. If the abstract path ∂Σ follows the primary path of current flow (marked in red), then the magnetic flux through this path changes dramatically as the plates are rotated, yet the EMF is almost zero.

One can analyze examples like these by taking care that the path $\partial\Sigma$ moves with the same velocity as the material. Alternatively, one can always correctly calculate the EMF by combining Lorentz force law with the Maxwell–Faraday equation:

$$\mathcal{E} = \int_{\partial\Sigma} (\mathbf{E} + \mathbf{v}_m \times \mathbf{B})\cdot d\mathbf{l} = -\int_{\Sigma} \frac{\partial \mathbf{B}}{\partial t} \cdot d\Sigma + \oint_{\partial\Sigma} (\mathbf{v}_m \times \mathbf{B})\cdot d\mathbf{l}$$

where "it is very important to notice that (1) [v_m] is the velocity of the conductor not the velocity of the path element dl and (2) in general, the partial derivative with respect to time cannot be moved outside the integral since the area is a function of time."

Faraday's Law and Relativity

Two Phenomena

Faraday's law is a single equation describing two different phenomena: the *motional EMF* generated by a magnetic force on a moving wire, and the *transformer EMF* generated by an electric force due to a changing magnetic field (described by the Maxwell–Faraday equation).

James Clerk Maxwell drew attention to this fact in his 1861 paper *On Physical Lines of Force*. In the latter half of Part II of that paper, Maxwell gives a separate physical explanation for each of the two phenomena.

A reference to these two aspects of electromagnetic induction is made in some modern textbooks. As Richard Feynman states:

So the "flux rule" that the emf in a circuit is equal to the rate of change of the magnetic flux through the circuit applies whether the flux changes because the field changes or because the circuit moves (or both).

Yet in our explanation of the rule we have used two completely distinct laws for the two cases – $v \times B$ for "circuit moves" and $\nabla \times E = -\partial_t B$ for "field changes".

We know of no other place in physics where such a simple and accurate general principle requires for its real understanding an analysis in terms of *two different phenomena*.

—Richard P. Feynman, The Feynman Lectures on Physics

Einstein's View

Reflection on this apparent dichotomy was one of the principal paths that led Einstein to develop special relativity:

It is known that Maxwell's electrodynamics—as usually understood at the present time—when applied to moving bodies, leads to asymmetries which do not appear to be inherent in the phenomena. Take, for example, the reciprocal electrodynamic action of a magnet and a conductor.

The observable phenomenon here depends only on the relative motion of the conductor and the magnet, whereas the customary view draws a sharp distinction between the two cases in which either the one or the other of these bodies is in motion. For if the magnet is in motion and the conductor at rest, there arises in the neighbourhood of the magnet an electric field with a certain definite energy, producing a current at the places where parts of the conductor are situated.

But if the magnet is stationary and the conductor in motion, no electric field arises in the neighbourhood of the magnet. In the conductor, however, we find an electromotive force, to which in itself there is no corresponding energy, but which gives rise—assuming equality of relative motion in the two cases discussed—to electric currents of the same path and intensity as those produced by the electric forces in the former case.

Examples of this sort, together with unsuccessful attempts to discover any motion of the earth relative to the "light medium," suggest that the phenomena of electrodynamics as well as of mechanics possess no properties corresponding to the idea of absolute rest.

—Albert Einstein, On the Electrodynamics of Moving Bodies

Lenz's Law

Lenz's law, named after the physicist Emil Lenz who formulated it in 1834, states that the direction of the current induced in a conductor by a changing magnetic field is such that the magnetic field created by the induced current opposes the initial changing magnetic field.

It is a qualitative law that specifies the direction of induced current, but states nothing about its magnitude. Lenz's law explains the direction of many effects in electromagnetism, such as the direction of voltage induced in an inductor or wire loop by a changing current, or the drag force of eddy currents exerted on moving objects in a magnetic field.

Lenz's law may be seen as analogous to Newton's third law in classic mechanics.

Lenz's law states that the current induced in a circuit due to a change or a motion in a magnetic field is so directed as to oppose the change in flux and to exert a mechanical force opposing the motion.

Lenz's law is contained in the rigorous treatment of Faraday's law of induction, where it finds expression by the negative sign:

$$\mathcal{E} = -\frac{\partial \Phi_B}{\partial t},$$

which indicates that the induced electromotive force $\mathcal{E}$ and the rate of change in magnetic flux Φ_B have opposite signs.

This means that the direction of the back EMF of an induced field opposes the changing current that is its cause. D.J. Griffiths summarized it as follows: *Nature abhors a change in flux.*

If a change in the magnetic field of current i_1 induces another electric current, i_2, the direction of i_2 is opposite that of the change in i_1. If these currents are in two coaxial circular conductors ℓ_1 and ℓ_2

respectively, and both are initially 0, then the currents i_1 and i_2 must counter-rotate. The opposing currents will repel each other as a result.

Example:

Currents bound inside the atoms of strong magnets can create counter-rotating currents in a copper or aluminum pipe. This is shown by dropping the magnet through the pipe. The descent of the magnet inside the pipe is observably slower than when dropped outside the pipe.

When a voltage is generated by a change in magnetic flux according to Faraday's law, the polarity of the induced voltage is such that it produces a current whose magnetic field opposes the change which produces it. The induced magnetic field inside any loop of wire always acts to keep the magnetic flux in the loop constant. In the examples below, if the flux is increasing, the induced field acts in opposition to it. If it is decreasing, the induced field acts in the direction of the applied field to oppose the change.

Detailed Interaction of Charges in these Currents

Experiment showing Lenz's law with two aluminium rings on a scales-like device set up on a pivot so as to freely move in the horizontal plane. One ring is fully enclosed, while the other has an opening, not forming a complete circle. When we place a bar magnet near the fully enclosed ring, the ring is repulsed by it. However, when the system comes to a rest, and we remove the bar magnet, then the ring is attracted by it. In the first case, the induced current created in the ring resists the increase of magnetic flux caused by the proximity of the magnet, while in the latter, taking the magnet out of the ring decreases the magnetic flux, inducing such current whose magnetic field resists the decrease of flux. This phenomenon is absent when we repeat the experiment with the ring that isn't enclosed by inserting and removing the magnet bar. The induced currents in this ring can't enclose themselves in the ring, and have a very weak field that cannot resist the change of the magnetic flux.

In electromagnetism, when charges move along electric field lines work is done on them, whether it involves storing potential energy (negative work) or increasing kinetic energy (positive work).

When net positive work is applied to a charge q_1, it gains speed and momentum. The net work on q_1 thereby generates a magnetic field whose strength (in units of magnetic flux density (1 tesla = 1 volt-second per square meter)) is proportional to the speed increase of q_1. This magnetic field can interact with a neighboring charge q_2, passing on this momentum to it, and in return, q_1 loses momentum.

The charge q_2 can also act on q_1 in a similar manner, by which it returns some of the momentum that it received from q_1. This back-and-forth component of momentum contributes to magnetic inductance. The closer that q_1 and q_2 are, the greater the effect. When q_2 is inside a conductive medium such as a thick slab made of copper or aluminum, it more readily responds to the force applied to it by q_1. The energy of q_1 is not instantly consumed as heat generated by the current of q_2 but is also stored in *two* opposing magnetic fields. The energy density of magnetic fields tends to vary with the square of the magnetic field's intensity; however, in the case of magnetically non-linear materials such as ferromagnets and superconductors, this relationship breaks down.

Conservation of Momentum

Momentum must be conserved in the process, so if q_1 is pushed in one direction, then q_2 ought to be pushed in the other direction by the same force at the same time. However, the situation becomes more complicated when the finite speed of electromagnetic wave propagation is introduced. This means that for a brief period the total momentum of the two charges is not conserved, implying that the difference should be accounted for by momentum in the fields, as asserted by Richard P. Feynman. Famous 19th century electrodynamicist James Clerk Maxwell called this the "electromagnetic momentum". Yet, such a treatment of fields may be necessary when Lenz's law is applied to opposite charges. It is normally assumed that the charges in question have the same sign. If they do not, such as a proton and an electron, the interaction is different. An electron generating a magnetic field would generate an EMF that causes a proton to accelerate in the same direction as the electron. At first, this might seem to violate the law of conservation of momentum, but such an interaction is seen to conserve momentum if the momentum of electromagnetic fields is taken into account.

Circuit Theorems

Kirchhoff's Laws

Kirchhoff's circuit laws are two equalities that deal with the current and potential difference (commonly known as voltage) in the lumped element model of electrical circuits. They were first described in 1845 by German physicist Gustav Kirchhoff. This generalized the work of Georg Ohm and preceded the work of James Clerk Maxwell. Widely used in electrical engineering, they are also called Kirchhoff's rules or simply Kirchhoff's laws. These laws can be applied in time and frequency domains and form the basis for network analysis.

Both of Kirchhoff's laws can be understood as corollaries of Maxwell's equations in the low-frequency limit. They are accurate for DC circuits, and for AC circuits at frequencies where the wavelengths of electromagnetic radiation are very large compared to the circuits.

Kirchhoff's Current Law

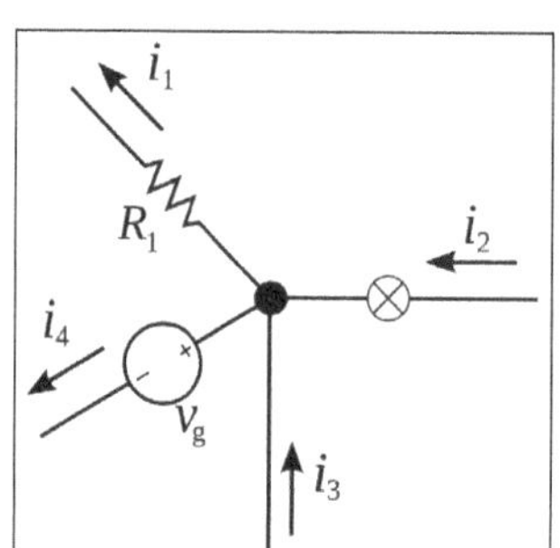

The current entering any junction is equal to the current leaving that junction. $i_2 + i_3 = i_1 + i_4$

This law is also called Kirchhoff's first law, Kirchhoff's point rule, or Kirchhoff's junction rule (or nodal rule).

This law states that, for any node (junction) in an electrical circuit, the sum of currents flowing into that node is equal to the sum of currents flowing out of that node; or equivalently:

The algebraic sum of currents in a network of conductors meeting at a point is zero.

Recalling that current is a signed (positive or negative) quantity reflecting direction towards or away from a node, this principle can be succinctly stated as:

$$\sum_{k=1}^{n} I_k = 0$$

where n is the total number of branches with currents flowing towards or away from the node.

The law is based on the conservation of charge where the charge (measured in coulombs) is the product of the current (in amperes) and the time (in seconds). If the net charge in a region is constant, the current law will hold on the boundaries of the region. This means that the current law relies on the fact that the net charge in the wires and components is constant.

Uses

A matrix version of Kirchhoff's current law is the basis of most circuit simulation software, such as SPICE. The current law is used with Ohm's law to perform nodal analysis.

The current law is applicable to any lumped network irrespective of the nature of the network; whether unilateral or bilateral, active or passive, linear or non-linear.

Kirchhoff's Voltage Law

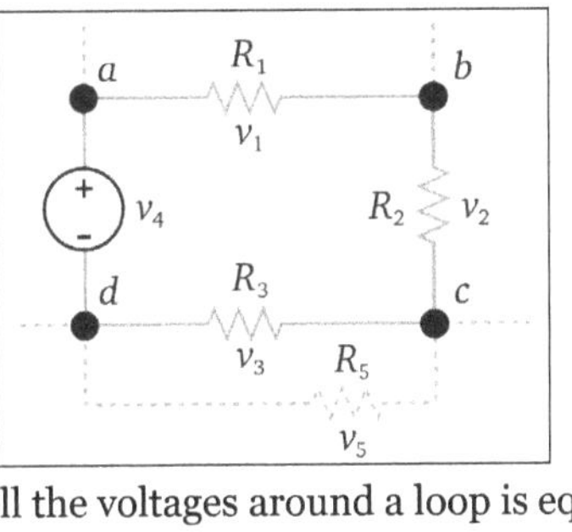

The sum of all the voltages around a loop is equal to zero.

$v_1 + v_2 + v_3 + v_4 = 0$

This law is also called Kirchhoff's second law, Kirchhoff's loop (or mesh) rule, and Kirchhoff's second rule.

This law states that:

> "The directed sum of the potential differences (voltages) around any closed loop is zero".

Similarly to Kirchhoff's current law, the voltage law can be stated as:

$$\sum_{k=1}^{n} V_k = 0$$

Here, n is the total number of voltages measured.

Derivation of Kirchhoff's Voltage Law

A similar derivation can be found in The Feynman Lectures on Physics, Volume II, Chapter 22: AC Circuits.

Consider some arbitrary circuit. Approximate the circuit with lumped elements, so that (time-varying) magnetic fields are contained to each component and the field in the region exterior to the circuit is negligible. Based on this assumption, the Maxwell-Faraday equation reveals that,

$$\nabla \times \mathbf{E} = -\frac{\partial \mathbf{B}}{\partial t} = \mathbf{0}$$

in the exterior region. If each of the components have a finite volume, then the exterior region is simply connected, and thus the electric field is conservative in that region. Therefore, for any loop in the circuit, we find that:

$$\sum V_i = -\sum \int_{\mathcal{P}_i} \mathbf{E} \cdot d\mathbf{l} = \oint \mathbf{E} \cdot d\mathbf{l} = 0$$

where $\mathcal{P}_i$ are paths around the exterior of each of the components, from one terminal to another.

Generalization

In the low-frequency limit, the voltage drop around any loop is zero. This includes imaginary loops arranged arbitrarily in space – not limited to the loops delineated by the circuit elements and conductors. In the low-frequency limit, this is a corollary of Faraday's law of induction (which is one of Maxwell's equations). This has practical application in situations involving "static electricity".

Limitations

Kirchhoff's circuit laws are the result of the lumped element model and both depend on the model being applicable to the circuit in question. When the model is not applicable, the laws do not apply.

The current law is dependent on the assumption that the net charge in any wire, junction or lumped component is constant. Whenever the electric field between parts of the circuit is non-negligible, such as when two wires are capacitively coupled, this may not be the case. This occurs in high-frequency AC circuits, where the lumped element model is no longer applicable. For example, in a transmission line, the charge density in the conductor will constantly be oscillating.

On the other hand, the voltage law relies on the fact that the action of time-varying magnetic fields are confined to individual components, such as inductors. In reality, the induced electric field produced by an inductor is not confined, but the leaked fields are often negligible.

Modelling Real Circuits with Lumped Elements

The lumped element approximation for a circuit is accurate at low frequencies. At higher frequencies, leaked fluxes and varying charge densities in conductors become significant. To an extent, it is possible to still model such circuits using parasitic components. If frequencies are too high, it may be more appropriate to simulate the fields directly using finite element modelling or other techniques.

To model circuits so that both laws can still be used, it is important to understand the distinction between *physical* circuit elements and the *ideal* lumped elements. For example, a wire is not an ideal conductor. Unlike an ideal conductor, wires can inductively and capacitively couple to each other (and to themselves), and have a finite propagation delay. Real conductors can be modeled in terms of lumped elements by considering parasitic capacitances distributed between the conductors to model capacitive coupling, or parasitic (mutual) inductances to model inductive coupling. Wires also have some self-inductance, which is the reason that decoupling capacitors are necessary.

Example:

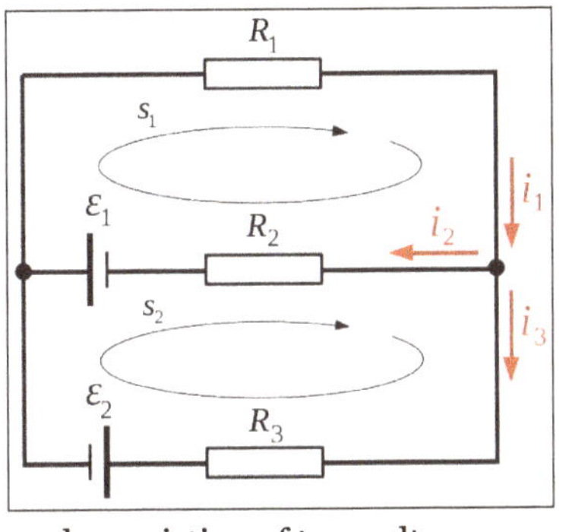

Assume an electric network consisting of two voltage sources and three resistors.

According to the first law:

$$i_1 - i_2 - i_3 = 0$$

Applying the second law to the closed circuit s_1, and substituting for voltage using Ohm's law gives:

$$-R_2 i_2 + \mathcal{E}_1 - R_1 i_1 = 0$$

The second law, again combined with Ohm's law, applied to the closed circuit s_2 gives:

$$-R_3 i_3 - \mathcal{E}_2 - \mathcal{E}_1 + R_2 i_2 = 0$$

This yields a system of linear equations in i_1, i_2, i_3:

$$\begin{cases} i_1 - i_2 - i_3 & = 0 \\ -R_2 i_2 + \mathcal{E}_1 - R_1 i_1 & = 0 \\ -R_3 i_3 - \mathcal{E}_2 - \mathcal{E}_1 + R_2 i_2 & = 0 \end{cases}$$

which is equivalent to,

$$\begin{cases} i_1 + (-i_2) + (-i_3) & = 0 \\ R_1 i_1 + R_2 i_2 + 0 i_3 & = \mathcal{E}_1 \\ 0 i_1 + R_2 i_2 - R_3 i_3 & = \mathcal{E}_1 + \mathcal{E}_2 \end{cases}$$

Assuming,

$$R_1 = 100\Omega,\ R_2 = 200\Omega,\ R_3 = 300\Omega$$
$$\mathcal{E}_1 = 3\text{V}, \mathcal{E}_2 = 4\text{V}$$

the solution is,

$$\begin{cases} i_1 = \dfrac{1}{1100}\text{A} \\ i_2 = \dfrac{4}{275}\text{A} \\ i_3 = -\dfrac{3}{220}\text{A} \end{cases}$$

The current i_3 has a negative sign which means the assumed direction of i_3 was incorrect and i_3 is actually flowing in the direction opposite to the red arrow labeled i_3. The current in R_3 flows from left to right.

Superposition Theorem

For every electrical circuit, there are two or additional independent supplies like current, voltage or both the sources. For examining these electrical circuits, superposition theoremis widely utilized and mostly for time-domain circuits at various frequencies. For instance, a linear DC circuit consists of one or more independent supply; we can get the supplies like voltage and current by using the methods like mesh analysis and nodal analysis techniques.

Otherwise, we can employ the "superposition theorem" that includes every individual supply result on the worth of the variable to be decided. This means the theorem assumes that every supply in a circuit independently discovers the rate of the variable, and lastly produces the secondary variable by inserting the variables which are reasoned by the effect of every source. Even though the process of it is very difficult but still can be applied for every linear circuit.

The superposition theorem is a method for the Independent supplies present in an electrical circuit like voltage & current and that is considered as one supply at a time. This theorem tells that in a linear n/w comprising one or more sources, the flow of current through a number of supplies in a circuit is the algebraic calculation of the currents when acting the sources like independently.

The application of this theorem involves simply in linear n/ws, and also in both the AC & DC circuits where it assists to build the circuits like "Norton" as well as "Thevenin" equivalent circuits.

For instance, the circuit which has two or more supplies then the circuit will be separated into a number of circuits based on the statement of the superposition theorem. Here, the separated circuits can make the entire circuit seem very simple in easier methods. And, by merging the separated circuits another time after individual circuit modification, one can simply discover factors like node voltages, voltage-drop at every resistance, currents, etc.

Step-by-Step Methods of Superposition Theorem Statement

The following step-by-step methods are used to discover the response of a circuit in a specific division by superposition theorem:

- Calculate the response in a specific branch of a circuit by allowing for one independent supply as well as removing the residual independent supplies current in the network.
- Do again the above step for all voltage and current sources there in the circuit.
- Include all the reactions in order to obtain the total response in a specific circuit when all the supplies are there in the network.

Superposition Theorem Examples

The basic circuit diagram of superposition theorem is shown below, and it is the best example of this theorem. By using this circuit, calculate the flow of current through the resistor R for the following circuit.

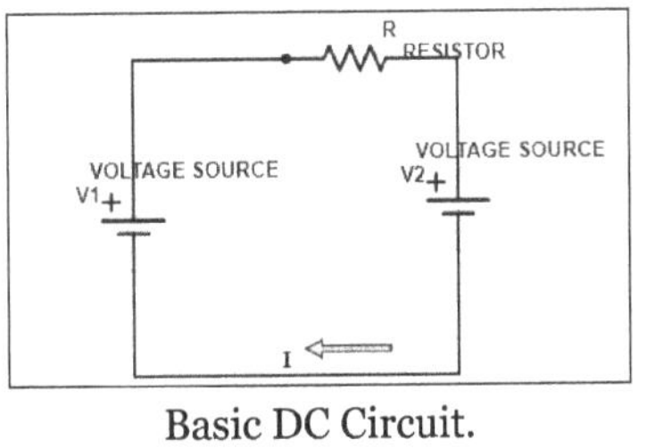

Basic DC Circuit.

Disable the secondary voltage source i.e, V2, and calculating the flow of current I1 in the following circuit.

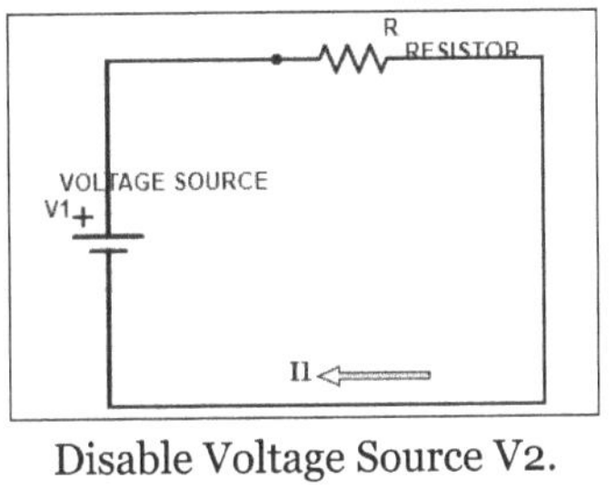

Disable Voltage Source V2.

We know that ohms law V= IR,

$I1 = V1/R$

Disable the primary voltage source i.e, V1, and calculating the flow of current I2 in the following circuit.

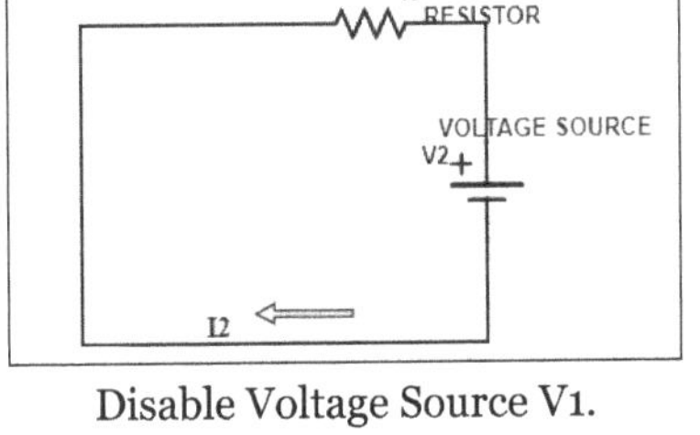

Disable Voltage Source V1.

I2= -V2/R

According to superposition theorem, the network current I = I1 + I2

I = V1/R-V2/R

Superposition Theorem Problems

The following circuit shows the basic DC circuit for solving the superposition theorem problem such that we can get the voltage across the load terminals. In the following circuit, there are two independent supplies namely current and voltage.

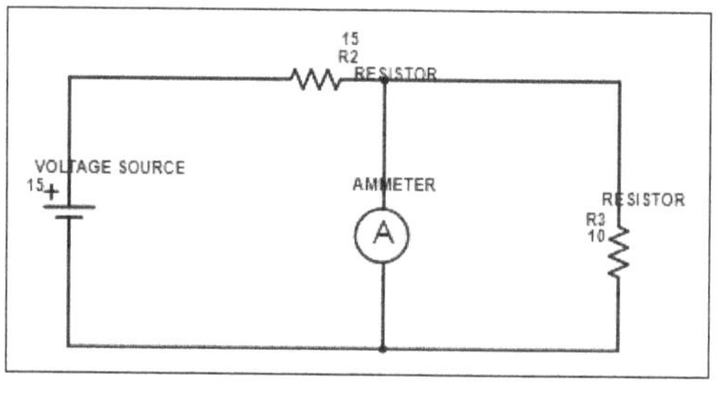

Initially, in the above circuit, we keep only voltage supply is acting, and the remaining supply like the current is changed with inside resistance. So the above circuit will become an open circuit as shown in the below figure.

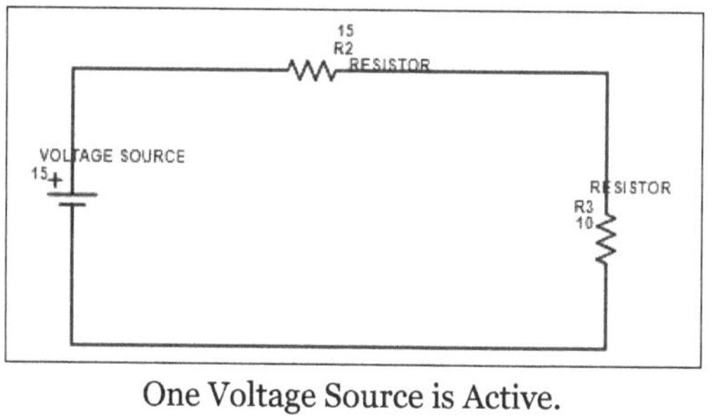

One Voltage Source is Active.

Consider the voltage across the load terminals VL1 with voltage supply performing alone, then

VL1 =Vs (R3/(R3 + R1))

Here, Vs= 15, R3= 10 and R2-= 15

Please substitute the above values in the above equation,

VL1 = Vs × R3 / (R3 + R2)

= 15 (10 / (10 + 15))

15(10/25)

= 6 Volts

Hold the current supply only and change the voltage supply with its inside resistance. So the circuit will become a short circuit as shown in the following figure.

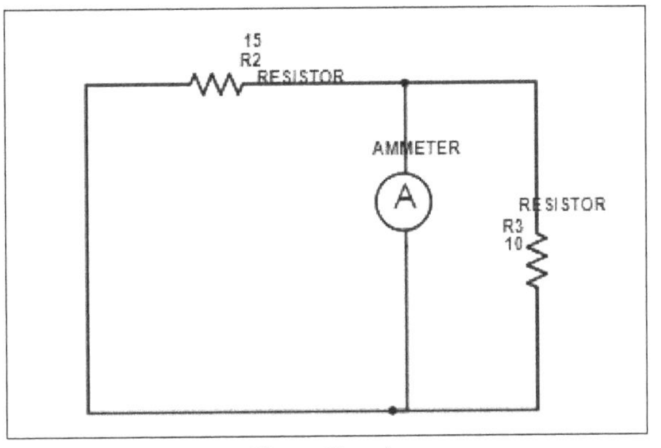

Short Circuit.

Consider the voltage across the load terminals is 'VL2' while only current supply performing. Then

VL2= I x R

IL = 1 x R1/(R1+R2)

R1 = 15 RL= 25

= 1 × 15 / (15 +25) = 0.375 Amps

VL2 = 0.375 × 10 = 3.75 Volts

As a result, we know that the superposition theorem states that the voltage across the load is the amount of VL1 & VL2

VL = VL1 + VL2

6 + 3.75 = 9.75 Volts

Applications and Limitations of Superposition Theorem

The superposition theorem cannot be useful for power calculations but this theorem works on the principle of linearity. As the power equation is not linear. As a result, the power used by the factor in a circuit with this theorem is not achievable.

If the load selection is changeable, then it is necessary to achieve each supply donation and their calculation for each transform in load resistance. So this is a very difficult method to analyze compound circuits.

The application of superposition theorem is, we can employ only for linear circuits as well as the circuit which has more supplies.

From the above superposition theorem examples, this theorem cannot be used for non-linear circuits, but applicable for linear circuits. The circuit can be examined with the single power source at a time, the equivalent section currents and voltages algebraically included discovering what they will perform with every power supplies in effect. To cancel out all except one power supply for study, substitute any power source with a cable; restore any current supply with the break.

Thevenin's Theorem

As originally stated in terms of DC resistive circuits only, Thévenin's theorem (aka Helmholtz–Thévenin theorem) holds that:

- Any linear electrical network containing only voltage sources, current sources and resistances can be replaced at terminals A-B by an equivalent combination of a voltage source V_{th} in a series connection with a resistance R_{th}.
- The equivalent voltage V_{th} is the voltage obtained at terminals A-B of the network with terminals A-B open circuited.
- The equivalent resistance R_{th} is the resistance that the circuit between terminals A and B would have if all ideal voltage sources in the circuit were replaced by a short circuit and all ideal current sources were replaced by an open circuit.
- If terminals A and B are connected to one another, the current flowing from A to B will be V_{th}/R_{th}. This means that R_{th} could alternatively be calculated as V_{th} divided by the short-circuit current between A and B when they are connected together.

In circuit theory terms, the theorem allows any one-port network to be reduced to a single voltage source and a single impedance.

The theorem also applies to frequency domain AC circuits consisting of reactive and resistive impedances. It means the theorem applies for AC in an exactly same way to DC except that resistances are generalized to impedances.

The theorem was independently derived in 1853 by the German scientist Hermann von Helmholtz and in 1883 by Léon Charles Thévenin (1857–1926), an electrical engineer with France's national Postes et Télégraphes telecommunications organization.

Thévenin's theorem and its dual, Norton's theorem, are widely used to make circuit analysis simpler and to study a circuit's initial-condition and steady-state response. Thévenin's theorem can be used to convert any circuit's sources and impedances to a Thévenin equivalent; use of the theorem may in some cases be more convenient than use of Kirchhoff's circuit laws.

Calculating the Thévenin equivalent

The equivalent circuit is a voltage source with voltage V_{Th} in series with a resistance R_{Th}.

The Thévenin-equivalent voltage V_{Th} is the open-circuit voltage at the output terminals of the original circuit. When calculating a Thévenin-equivalent voltage, the voltage divider principle is often useful, by declaring one terminal to be V_{out} and the other terminal to be at the ground point.

The Thévenin-equivalent resistance R_{Th} is the resistance measured across points A and B "looking back" into the circuit. The resistance is measured after replacing all voltage- and current-sources with their internal resistances. That means an ideal voltage source is replaced with a short circuit, and an ideal current source is replaced with an open circuit. Resistance can then be calculated across the terminals using the formulae for series and parallel circuits. This method is valid only for circuits with independent sources. If there are dependent sources in the circuit, another method must be used such as connecting a test source across A and B and calculating the voltage across or current through the test source.

The replacements of voltage and current sources do what the sources would do if their values were set to zero. A zero valued voltage source would create a potential difference of zero volts between its terminals, regardless of the current that passes through it; its replacement, a short circuit, does the same thing. A zero valued current source passes zero current, regardless of the voltage across it; its replacement, an open circuit, does the same thing.

Example:

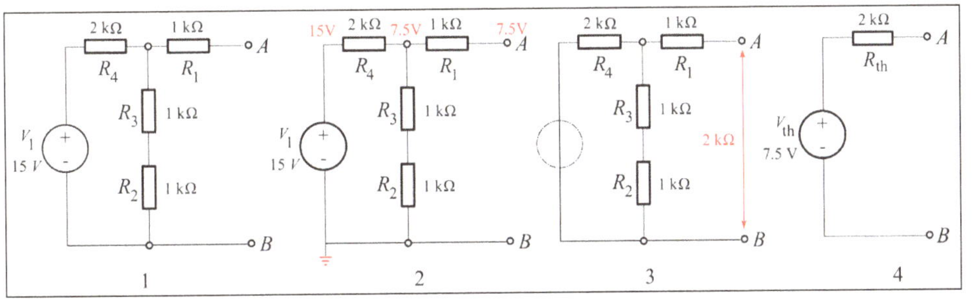

- Original circuit.
- The equivalent voltage.
- The equivalent resistance.
- The equivalent circuit.

In the example, calculating the equivalent voltage:

$$V_{\mathrm{Th}} = \frac{R_2 + R_3}{(R_2 + R_3) + R_4} \cdot V_1$$
$$= \frac{1\,\mathrm{k\Omega} + 1\,\mathrm{k\Omega}}{(1\,\mathrm{k\Omega} + 1\,\mathrm{k\Omega}) + 2\,\mathrm{k\Omega}} \cdot 15\,\mathrm{V}$$
$$= \frac{1}{2} \cdot 15\,\mathrm{V} = 7.5\,\mathrm{V}$$

Notice that R_1 is not taken into consideration, as above calculations are done in an open-circuit condition between A and B, therefore no current flows through this part, which means there is no current through R_1 and therefore no voltage drop along this part.

Calculating equivalent resistance ($R_x \mid R_y$) is the total resistance of two parallel resistors):

$$R_{\mathrm{Th}} = R_1 + \left[\left(R_2 + R_3\right) \| R_4\right]$$

$$= 1\,\mathrm{k\Omega} + \left[\left(1\,\mathrm{k\Omega} + 1\,\mathrm{k\Omega}\right) \| 2\,\mathrm{k\Omega}\right]$$

$$= 1\,\mathrm{k\Omega} + \left(\frac{1}{(1\,\mathrm{k\Omega} + 1\,\mathrm{k\Omega})} + \frac{1}{(2\,\mathrm{k\Omega})}\right)^{-1} = 2\,\mathrm{k\Omega}.$$

Conversion to a Norton Equivalent

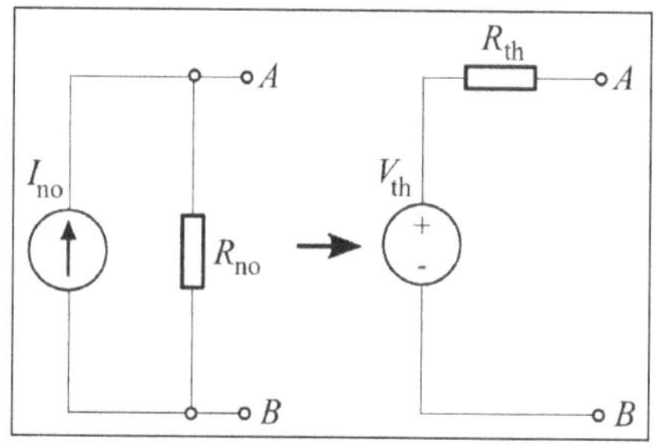

Norton-Thevenin conversion.

A Norton equivalent circuit is related to the Thévenin equivalent by,

$$R_{\text{Th}} = R_{\text{No}}$$

$$V_{\text{Th}} = I_{\text{No}} R_{\text{No}}$$

$$I_{\text{No}} = V_{\text{Th}} / R_{\text{Th}}$$

Practical Limitations

- Many circuits are only linear over a certain range of values, thus the Thévenin equivalent is valid only within this linear range.
- The Thévenin equivalent has an equivalent I–V characteristic only from the point of view of the load.
- The power dissipation of the Thévenin equivalent is not necessarily identical to the power dissipation of the real system. However, the power dissipated by an external resistor between the two output terminals is the same regardless of how the internal circuit is implemented.

A Proof of the Theorem

The proof involves two steps. The first step is to use superposition theorem to construct a solution. Then, uniqueness theorem is employed to show that the obtained solution is unique. It is noted that the second step is usually implied in literature.

By using superposition of specific configurations, it can be shown that for any linear "black box" circuit which contains voltage sources and resistors, its voltage is a linear function of the corresponding current as follows:

$$V = V_{\text{Eq}} - Z_{\text{Eq}} I.$$

Here, the first term reflects the linear summation of contributions from each voltage source, while the second term measures the contributions from all the resistors. The above expression is obtained by using the fact that the voltage of the black box for a given current I is identical to the linear superposition of the solutions of the following problems: (1) to leave the black box open circuited but activate individual voltage source one at a time and, (2) to short circuit all the voltage sources but feed the circuit with a certain ideal voltage source so that the resulting current exactly reads I (Alternatively, one can use an ideal current source of current I). Moreover, it is straightforward to show that V_{Eq} and Z_{Eq} are the single voltage source and the single series resistor in question.

As a matter of fact, the above relation between V and I is established by superposition of some particular configurations. Now, the uniqueness theorem guarantees that the result is general. To be specific, there is one and only one value of V once the value of I is given. In other words, the above relation holds true independent of what the "black box" is plugged to.

Norton's Theorem

Norton's theorem (aka Mayer–Norton theorem) holds, to illustrate in DC circuit theory terms:

- Any linear electrical network with voltage and current sources and only resistances can be replaced at terminals A–B by an equivalent current source Ino in parallel connection with an equivalent resistance Rno.
- This equivalent current Ino is the current obtained at terminals A-B of the network with terminals A-B short circuited.
- This equivalent resistance Rno is the resistance obtained at terminals A-B of the network with all its voltage sources short circuited and all its current sources open circuited.

For alternating current (AC) systems the theorem can be applied to reactive impedances as well as resistances.

The Norton equivalent circuit is used to represent any network of linear sources and impedances at a given frequency.

Norton's theorem and its dual, Thévenin's theorem, are widely used for circuit analysis simplification and to study circuit's initial-condition and steady-state response.

Norton's theorem was independently derived in 1926 by Siemens & Halske researcher Hans Ferdinand Mayer and Bell Labs engineer Edward Lawry Norton.

To find the equivalent,

- Find the Norton current I_{no}. Calculate the output current, I_{AB}, with a short circuit as the load (meaning 0 resistance between A and B). This is I_{no}.
- Find the Norton resistance R_{no}. When there are no dependent sources (all current and voltage sources are independent), there are two methods of determining the Norton impedance R_{no}.
    - Calculate the output voltage, V_{AB}, when in open circuit condition (i.e., no load resistor – meaning infinite load resistance). R_{no} equals this V_{AB} divided by I_{no}.

or

- ◦ Replace independent voltage sources with short circuits and independent current sources with open circuits. The total resistance across the output port is the Norton impedance R_{no}.

This is equivalent to calculating the Thevenin resistance.

However, when there are dependent sources, the more general method must be used. This method is not shown below in the diagrams.

- Connect a constant current source at the output terminals of the circuit with a value of 1 ampere and calculate the voltage at its terminals. This voltage divided by the 1 A current is the Norton impedance R_{no}. This method must be used if the circuit contains dependent sources, but it can be used in all cases even when there are no dependent sources.

Example of a Norton Equivalent Circuit

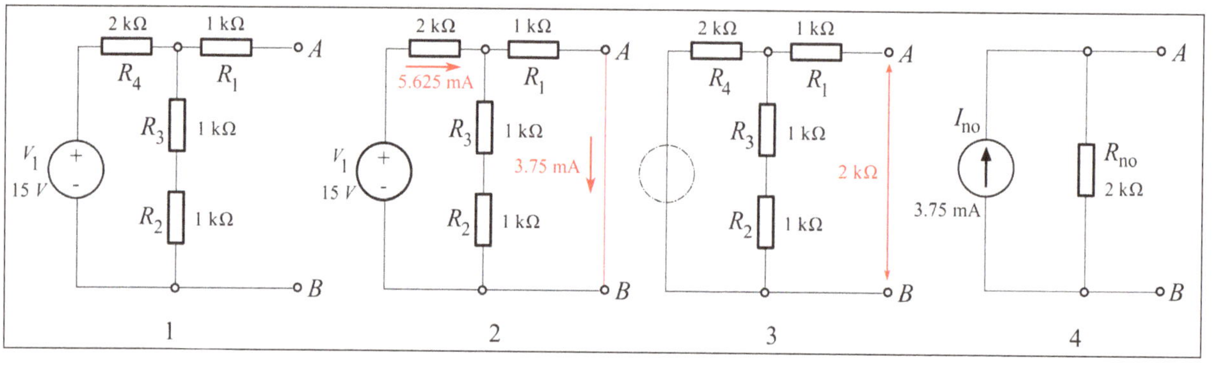

- The original circuit.
- Calculating the equivalent output current.
- Calculating the equivalent resistance.
- Design the Norton equivalent circuit.

In the example, the total current I_{total} is given by:

$$I_{total} = \frac{15\,\text{V}}{2\,\text{k}\Omega + 1\,\text{k}\Omega \backslash (1\,\text{k}\Omega + 1\,\text{k}\Omega)} = 5.625\,\text{mA}.$$

The current through the load is then, using the current divider rule:

$$I_{no} = \frac{1\,\text{k}\Omega + 1\,\text{k}\Omega}{(1\,\text{k}\Omega + 1\,\text{k}\Omega + 1\,\text{k}\Omega)} \cdot I_{total}$$

$$= 2/3 \cdot 5.625\,\text{mA} = 3.75\,\text{mA}.$$

And the equivalent resistance looking back into the circuit is:

$$R_{no} = 1\,k\Omega + (2\,k\Omega \backslash (1\,k\Omega + 1\,k\Omega)) = 2\,k\Omega.$$

So the equivalent circuit is a 3.75 mA current source in parallel with a 2 kΩ resistor.

Conversion to a Thévenin Equivalent

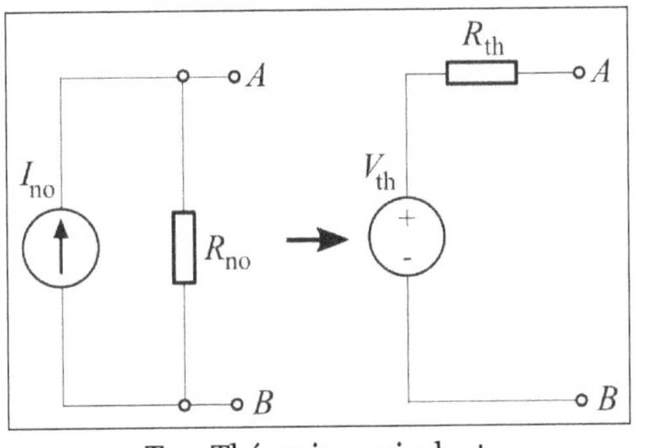

To a Thévenin equivalent.

A Norton equivalent circuit is related to the Thévenin equivalent by the equations:

$$R_{th} = R_{no}$$

$$V_{th} = I_{no} R_{no}$$

$$\frac{V_{th}}{R_{th}} = I_{no}$$

Queueing Theory

The passive circuit equivalent of "Norton's theorem" in queuing theory is called the Chandy Herzog Woo theorem. In a reversible queueing system, it is often possible to replace an uninteresting subset of queues by a single (FCFS or PS) queue with an appropriately chosen service rate.

Maximum Power Transfer Theorem

Maximum Power Transfer occurs when the resistive value of the load is equal in value to that of the voltage sources internal resistance allowing maximum power to be supplied. Generally, this source resistance or even impedance if inductors or capacitors are involved is of a fixed value in Ohm's.

However, when we connect a load resistance, R_L across the output terminals of the power source, the impedance of the load will vary from an open-circuit state to a short-circuit state resulting in the power being absorbed by the load becoming dependent on the impedance of the actual power source. Then for the load resistance to absorb the maximum power possible it has to be "Matched" to the impedance of the power source and this forms the basis of Maximum Power Transfer.

The Maximum Power Transfer Theorem is another useful circuit analysis method to ensure that the maximum amount of power will be dissipated in the load resistance when the value of the load resistance is exactly equal to the resistance of the power source. The relationship between the load impedance and the internal impedance of the energy source will give the power in the load. Consider the circuit below.

Thevenins Equivalent Circuit

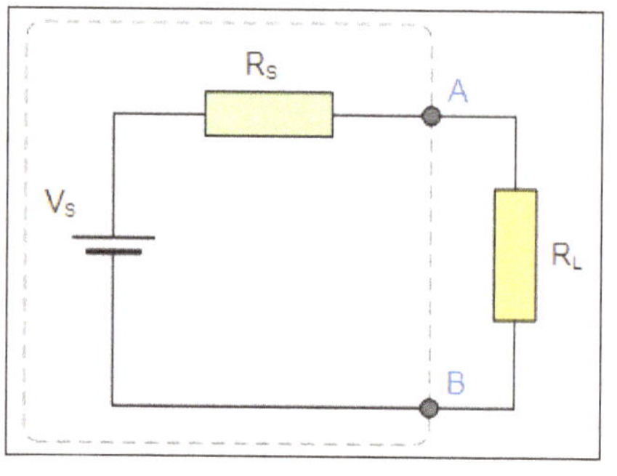

In our Thevenin equivalent circuit above, the maximum power transfer theorem states that "the maximum amount of power will be dissipated in the load resistance if it is equal in value to the Thevenin or Norton source resistance of the network supplying the power".

In other words, the load resistance resulting in greatest power dissipation must be equal in value to the equivalent Thevenin source resistance, then $R_L = R_S$ but if the load resistance is lower or higher in value than the Thevenin source resistance of the network, its dissipated power will be less than maximum.

For example, find the value of the load resistance, R_L that will give the maximum power transfer in the following circuit.

Maximum Power Transfer Example

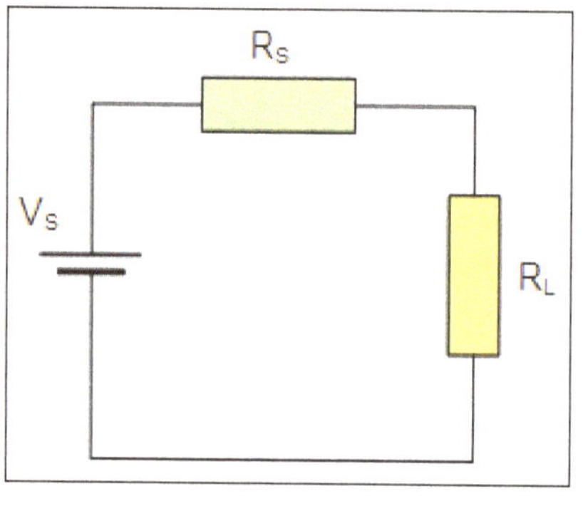

Where:

$R_S = 25\Omega$

R_L is variable between 0 – 100Ω

V_S = 100v

Then by using the following Ohm's Law equations:

$$I = \frac{V_S}{R_S + R_L} \text{ and } P = I^2 R_L$$

We can now complete the following table to determine the current and power in the circuit for different values of load resistance.

Table: Table of current against power.

R_L (Ω)	I (amps)	P (watts)	R_L (Ω)	I (amps)	P (watts)
0	4.0	0	25	2.0	100
5	3.3	55	30	1.8	97
10	2.8	78	40	1.5	94
15	2.5	93	60	1.2	83
20	2.2	97	100	0.8	64

Using the data from the table above, we can plot a graph of load resistance, R_L against power, P for different values of load resistance. Also notice that power is zero for an open-circuit (zero current condition) and also for a short-circuit (zero voltage condition).

Graph of Power against Load Resistance

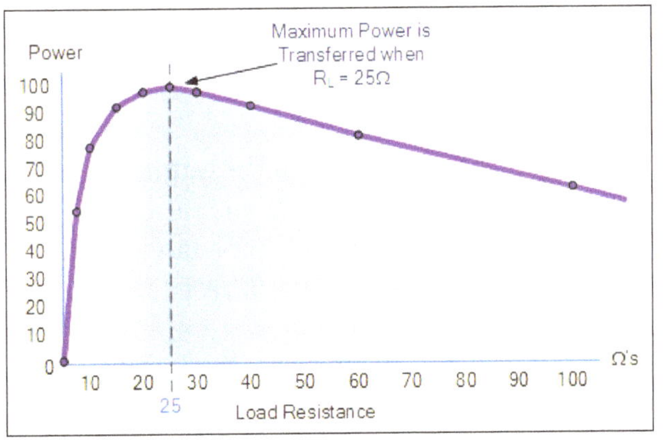

From the above table and graph we can see that the Maximum Power Transferoccurs in the load when the load resistance, R_L is equal in value to the source resistance, R_S that is: $R_S = R_L$ = 25Ω. This is called a "matched condition" and as a general rule, maximum power is transferred from an active device such as a power supply or battery to an external device when the impedance of the external device exactly matches the impedance of the source.

One good example of impedance matching is between an audio amplifier and a loudspeaker. The output impedance, Z_{OUT} of the amplifier may be given as between 4Ω and 8Ω, while the nominal input impedance, Z_{IN} of the loudspeaker may be given as 8Ω only.

Then if the 8Ω speaker is attached to the amplifiers output, the amplifier will see the speaker as an 8Ω load. Connecting two 8Ω speakers in parallel is equivalent to the amplifier driving one 4Ω speaker and both configurations are within the output specifications of the amplifier.

Improper impedance matching can lead to excessive power loss and heat dissipation. But how could you impedance match an amplifier and loudspeaker which have very different impedances. Well, there are loudspeaker impedance matching transformers available that can change impedances from 4Ω to 8Ω, or to 16Ω's to allow impedance matching of many loudspeakers connected together in various combinations such as in PA (public address) systems.

Transformer Impedance Matching

One very useful application of impedance matching in order to provide maximum power transfer between the source and the load is in the output stages of amplifier circuits. Signal transformers are used to match the loudspeakers higher or lower impedance value to the amplifiers output impedance to obtain maximum sound power output. These audio signal transformers are called "matching transformers" and couple the load to the amplifiers output as shown below.

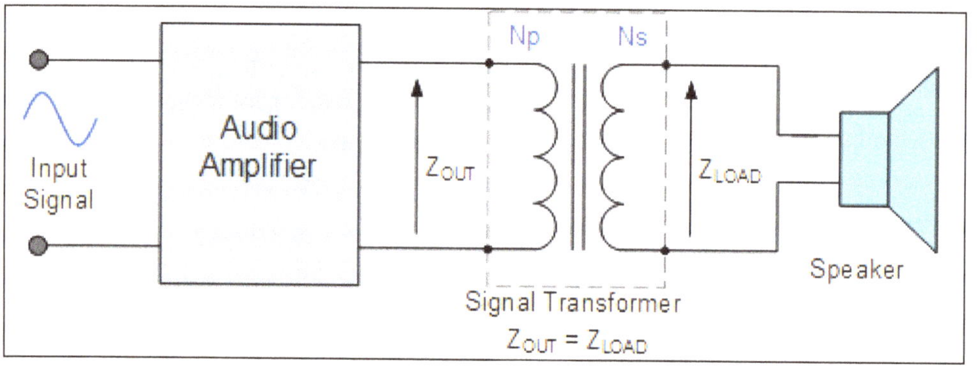

The maximum power transfer can be obtained even if the output impedance is not the same as the load impedance. This can be done using a suitable "turns ratio" on the transformer with the corresponding ratio of load impedance, Z_{LOAD} to output impedance, Z_{OUT} matches that of the ratio of the transformers primary turns to secondary turns as a resistance on one side of the transformer becomes a different value on the other.

If the load impedance, Z_{LOAD} is purely resistive and the source impedance is purely resistive, Z_{OUT} then the equation for finding the maximum power transfer is given as:

$$Z_{out} = \left(\frac{N_P}{N_S}\right)^2 Z_{load}$$

Where: N_P is the number of primary turns and N_S the number of secondary turns on the transformer. Then by varying the value of the transformers turns ratio the output impedance can be "matched" to the source impedance to achieve maximum power transfer.

Maximum Power Transfer Example

If an 8Ω loudspeaker is to be connected to an amplifier with an output impedance of 1000Ω, calculate the turns ratio of the matching transformer required to provide maximum power transfer of the audio signal. Assume the amplifier source impedance is Z_1, the load impedance is Z_2 with the turns ratio given as N.

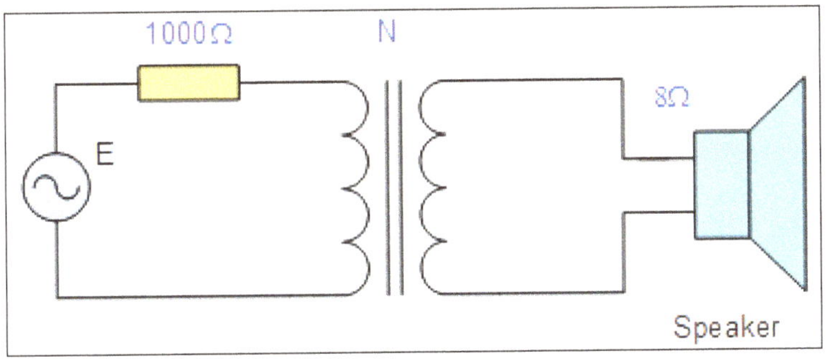

$$Z_1 = N^2 Z_2 \therefore N = \sqrt{\frac{Z_1}{Z_2}}$$

therefore,

$$N = \sqrt{\frac{Z_1}{Z_2}} = \sqrt{\frac{1000}{8}} = 11:2:1$$

Generally, small high frequency audio transformers used in low power amplifier circuits are nearly always regarded as ideal for simplicity, so any losses can be ignored.

References

- Dorf, Richard C.; Svoboda, James A. (2010). "Chapter 5: Circuit Theorems". Introduction to Electric Circuits (8th ed.). Hoboken, NJ, USA: John Wiley & Sons. Pp. 162–207. ISBN 978-0-470-52157-1. Archived from the original on 2012-04-30. Retrieved 2018-12-08
- Ohms-law-and-resistance, electricity, physics, guides: toppr.com, Retrieved 30 January, 2019
- Knoepfel, Heinz E. (2000). Magnetic Fields: A comprehensive theoretical treatise for practical use. Wiley. P. 4. ISBN 0-471-32205-9
- Superposition-theorem-example-problems-limitations-and-applications: elprocus.com, Retrieved 31 March, 2019
- Grant, I. S.; Phillips, W. R. (2008). Electromagnetism. Manchester Physics (2nd ed.). John Wiley & Sons. ISBN 978-0-471-92712-9
- Dcp-9, dccircuits: tutorials.ws, Retrieved 31 May, 2019
- Tipler, Paul A.; Mosca, Gene (2008). Physics for Scientists and Engineers (6th ed.). New York: W. H. Freeman and Company. ISBN 978-0-7167-8964-2. LCCN 2007010418

5

Tools and Instruments in Electrical Engineering

Various types of devices and equipment are used in electrical engineering. Some of them are LCR meter, voltmeter, ohmmeter, ammeter, current clamp, multimeter, network analyzer, power factor meter, etc. This chapter has been carefully written to provide an easy understanding of these tools and instruments used in electrical engineering.

LCR Meter

An LCR meter is used to measure the impedance of a circuit or a device. The LCR meter also measures D or Q. D stands for the dissipation factor. It is given by dividing the real part of the impedance by the imaginary part of the impedance (which is the reactance). Q stands for the quality factor. It is the inverse of D.

An LCR meter is used to measure the inductance, capacitance and resistance of a circuit. Hence, the name LCR meter.

When there is a change in the current flowing through a conductor, a corresponding change is induced in the voltage in it and in conductors surrounding it. This property is known as inductance. The ability of a body or a conductor to store electrical charge is known as capacitance. The opposition that a conductor offers to the passage of electric current through it is called resistance.

Working Principle

We pass an AC voltage through a DUT (Device Under Test). Now the LCR meter is used to measure the voltage and the current across the DUT. The magnitude of the impedance can be calculated from the ratio of these two quantities.

Uses of an LCR Meter

A digital LCR meter is used to measure the impedance flowing through a Device Under Test (DUT). It measures the voltage (V) across it, the current (I) flowing through it, and the phase angle between current and voltage. Subsequently, we can determine all the impedance parameters from these three factors.

Thus, an LCR meter measures the following parameters related to a circuit:

- Inductance,
- Capacitance,
- Resistance,
- Dissipation factor,
- Quality factor,
- Current,
- Voltage,
- Phase angle between the current and voltage,
- Conductance,
- Susceptance.

What is Impedance?

Impedance is the opposition that a circuit offers to the flow of direct or alternating current through it. It is a vector quantity composed of 2 scalars: resistance and reactance. Reactance is the name given to the opposition by an electronic component to the flow of alternating current due to capacitance and inductance.

Types of LCR Meters

Handheld LCR Meters

As the name suggests, these LCR meters are small in size and can be held in the hand; they are lightweight and also portable. They have a multiple test frequency, and the data it captures can be transferred to a PC via a USB port. They are generally used in field operations. They offer an accuracy in the range of 0.2% to 0.1%. The test frequency of a handheld LCR meter varies from 100 Hz, 120 Hz, 1 kHz, 10 kHz, and 100 kHz.

Benchtop LCR Meters

They are bulky in size. They can be operated on programmable frequencies. They offer an accuracy of 0.01%. They can be controlled via a computer. Test frequencies are above 100kHz.

Test Frequency

LCR meters operate within the frequency range from 10Hz to 2Mhz. The DUT is used under its own frequency. LCR meters have to match the frequency of their measurement to the frequency that the DUT is tuned to.

Test Voltage

A particular voltage is applied to the DUT. The AC output voltage of an LCR meter has to match it.

Accuracy

If the accuracy of your measurement is high, it takes more time to record that measurement for the LCR meter. Accuracy is hampered if the measurement is recorded in a short time. Most LCR meters provide 3 speeds for measurement: slow, medium, and fast. You have to make a choice between speed and accuracy.

Some Techniques used with LCR Meters

Bridge Method

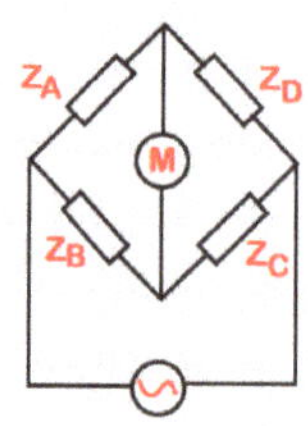

LCR meter in a wheatstone bridge.

This method is employed for measuring frequencies below 100kHz. The DUT is placed in a Wheatstone Bridge. Z_D is the DUT. Z_B and Z_C are known impedances. Impedance of Z_A is varied until no current flows through Z_D.

Thus, the four impedances obey the equation:

$$Z_D / Z_A = Z_C / Z_B$$

$$Z_D = (Z_C / Z_B) Z_A$$

This way, we can find the impedance of the DUT.

LCR Measurement by Current-Voltage Technique

In this technique, the LCR measurement of a component is done by measuring the current and voltage. Then the impedance values are found out from these two quantities.

There are different arrangements for low impedance and high impedance circuits, which are as follows:

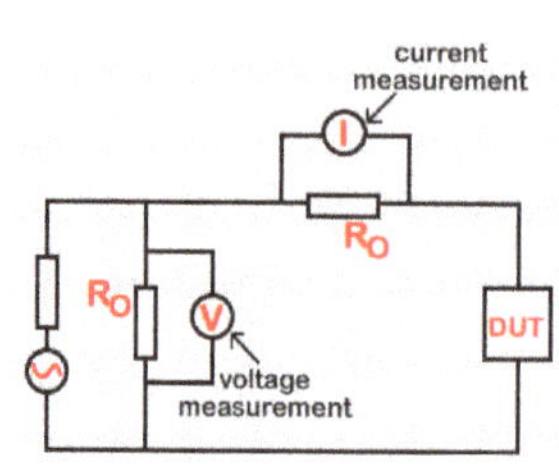

LCR measurement by current and voltage technique for high impedance current.

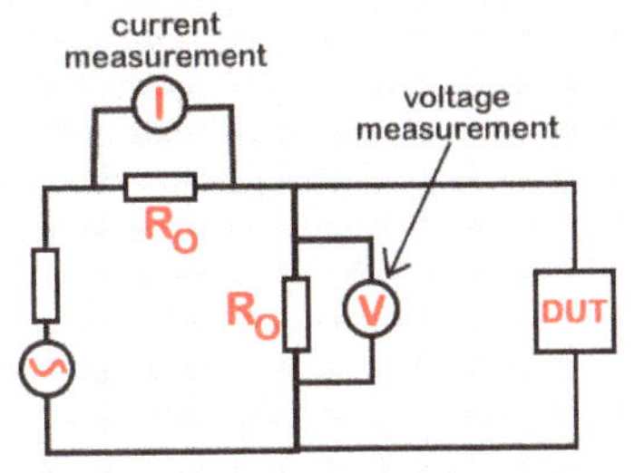

LCR measurement by current and voltage technique for low impedance current.

Both analog and digital LCR meters are available. While analog testers are cheaper, the digital variety scores on quality as it is more accurate.

Voltmeter

A voltmeter is an instrument used for measuring electrical potential difference between two points in an electric circuit. Analog voltmeters move a pointer across a scale in proportion to the voltage of the circuit; digital voltmeters give a numerical display of voltage by use of an analog to digital converter.

A voltmeter in a circuit diagram is represented by the letter *V* in a circle.

Voltmeters are made in a wide range of styles. Instruments permanently mounted in a panel are used to monitor generators or other fixed apparatus. Portable instruments, usually equipped to also measure current and resistance in the form of a multimeter, are standard test instruments used in electrical and electronics work. Any measurement that can be converted to a voltage can be displayed on a meter that is suitably calibrated; for example, pressure, temperature, flow or level in a chemical process plant.

General purpose analog voltmeters may have an accuracy of a few percent of full scale, and are used with voltages from a fraction of a volt to several thousand volts. Digital meters can be made with high accuracy, typically better than 1%. Specially calibrated test instruments have higher accuracies, with laboratory instruments capable of measuring to accuracies of a few parts per million. Meters using amplifiers can measure tiny voltages of microvolts or less.

Part of the problem of making an accurate voltmeter is that of calibration to check its accuracy. In laboratories, the Weston Cell is used as a standard voltage for precision work. Precision voltage references are available based on electronic circuits.

Analog Voltmeter

A moving coil galvanometer can be used as a voltmeter by inserting a resistor in series with the instrument. The galvanometer has a coil of fine wire suspended in a strong magnetic field. When an electric current is applied, the interaction of the magnetic field of the coil and of the stationary magnet creates a torque, tending to make the coil rotate. The torque is proportional to the current through the coil. The coil rotates, compressing a spring that opposes the rotation. The deflection of the coil is thus proportional to the current, which in turn is proportional to the applied voltage, which is indicated by a pointer on a scale.

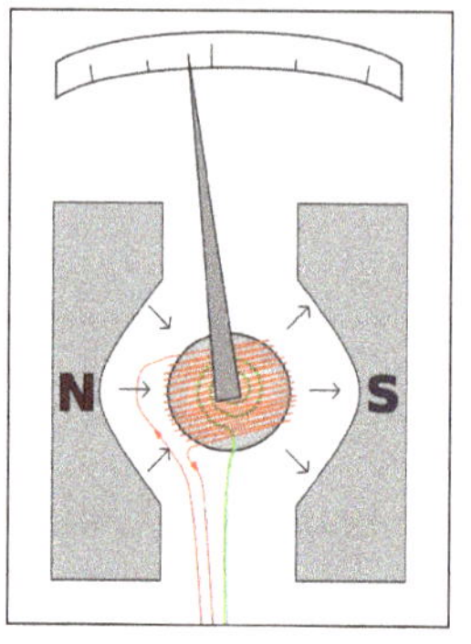

A moving coil galvanometer of the d'Arsonval type:

- The red wire carries the current to be measured.
- The restoring spring is shown in green.
- N and S are the north and south poles of the magnet.

One of the design objectives of the instrument is to disturb the circuit as little as possible and so the instrument should draw a minimum of current to operate. This is achieved by using a sensitive galvanometer in series with a high resistance, and then the entire instrument is connected in parallel with the circuit examined.

The sensitivity of such a meter can be expressed as "ohms per volt", the number of ohms resistance in the meter circuit divided by the full scale measured value. For example, a meter with a sensitivity of 1000 ohms per volt would draw 1 milliampere at full scale voltage; if the full scale was 200 volts, the resistance at the instrument's terminals would be 200000 ohms and at full scale the meter would draw 1 milliampere from the circuit under test. For multi-range instruments, the input resistance varies as the instrument is switched to different ranges.

Moving-coil instruments with a permanent-magnet field respond only to direct current. Measurement of AC voltage requires a rectifier in the circuit so that the coil deflects in only one direction. Some moving-coil instruments are also made with the zero position in the middle of the scale instead of at one end; these are useful if the voltage reverses its polarity.

Voltmeters operating on the electrostatic principle use the mutual repulsion between two charged plates to deflect a pointer attached to a spring. Meters of this type draw negligible current but are sensitive to voltages over about 100 volts and work with either alternating or direct current.

VTVMs and FET-VMs

The sensitivity and input resistance of a voltmeter can be increased if the current required to deflect the meter pointer is supplied by an amplifier and power supply instead of by the circuit under test. The electronic amplifier between input and meter gives two benefits; a rugged moving coil instrument can be used, since its sensitivity need not be high, and the input resistance can be made high, reducing the current drawn from the circuit under test. Amplified voltmeters often have an input resistance of 1, 10, or 20 megohms which is independent of the range selected. A once-popular form of this instrument used a vacuum tube in the amplifier circuit and so was called the vacuum tube voltmeter, or VTVM. These were almost always powered by the local AC line current and so were not particularly portable. Today these circuits use a solid-state amplifier using field-effect transistors, hence FET-VM, and appear in handheld digital multimeters as well as in bench and laboratory instruments. These are now so ubiquitous that they have largely replaced non-amplified multimeters except in the least expensive price ranges.

Most VTVMs and FET-VMs handle DC voltage, AC voltage, and resistance measurements; modern FET-VMs add current measurements and often other functions as well. A specialized form of the VTVM or FET-VM is the AC voltmeter. These instruments are optimized for measuring AC voltage. They have much wider bandwidth and better sensitivity than a typical multifunction device.

Digital Voltmeter

Two digital voltmeters. Note the 40 microvolt difference between the two measurements, an offset of 34 parts per million.

A digital voltmeter (DVM) measures an unknown input voltage by converting the voltage to a digital value and then displays the voltage in numeric form. DVMs are usually designed around a special type of analog-to-digital converter called an integrating converter.

DVM measurement accuracy is affected by many factors, including temperature, input impedance, and DVM power supply voltage variations. Less expensive DVMs often have input resistance on the order of 10 MΩ. Precision DVMs can have input resistances of 1 GΩ or higher for the lower voltage ranges (e.g. less than 20 V). To ensure that a DVM's accuracy is within the manufacturer's specified tolerances, it must be periodically calibrated against a voltage standard such as the Weston cell.

The first digital voltmeter was invented and produced by Andrew Kay of Non-Linear Systems (and later founder of Kaypro) in 1954.

Ohmmeter

An ohmmeter is an electrical instrument that measures electrical resistance, the opposition to an electric current. Micro-ohmmeters (microhmmeter or micro ohmmeter) make low resistance measurements. Megohmmeters (also a trademarked device Megger) measure large values of resistance. The unit of measurement for resistance is ohms (Ω).

Design Evolution

The first ohmmeters were based on a type of meter movement known as a 'ratiometer'. These were similar to the galvanometer type movement encountered in later instruments, but instead of hair-springs to supply a restoring force they used conducting 'ligaments'. These provided no net rotational force to the movement. Also, the movement was wound with two coils. One was connected via a series resistor to the battery supply. The second was connected to the same battery supply via a second resistor and the resistor under test. The indication on the meter was proportional to the ratio of the currents through the two coils. This ratio was determined by the magnitude of the resistor under test. The advantages of this arrangement were twofold. First, the indication of the resistance was completely independent of the battery voltage (as long as it actually produced some voltage) and no zero adjustment was required. Second, although the resistance scale was non linear, the scale remained correct over the full deflection range. By interchanging the two coils a second range was provided. This scale was reversed compared to the first. A feature of this type of instrument was that it would continue to indicate a random resistance value once the test leads were disconnected (the action of which disconnected the battery from the movement). Ohmmeters of this type only ever measured resistance as they could not easily be incorporated into a multimeter design. Insulation testers that relied on a hand cranked generator operated on the same principle. This ensured that the indication was wholly independent of the voltage actually produced.

Subsequent designs of ohmmeter provided a small battery to apply a voltage to a resistance via a galvanometer to measure the current through the resistance (battery, galvanometer and resistance all connected in series). The scale of the galvanometer was marked in ohms, because the fixed voltage from the battery assured that as resistance is increased, the current through the meter (and hence deflection) would decreased. Ohmmeters form circuits by themselves, therefore they cannot be used within an assembled circuit. This design is much simpler and cheaper than the former design, and was simple to integrate into a multimeter design and consequently was by far the most common form of analogue ohmmeter. This type of ohmmeter suffers from two inherent disadvantages. First, the meter needs to be zeroed by shorting the measurement points together and performing an adjustment for zero ohms indication prior to each measurement. This is because as the battery voltage decreases with age, the series resistance in the meter needs to be reduced to maintain the zero indication at full deflection. Second, and consequent on the first, the actual deflection for any given resistor under test changes as the internal resistance is altered. It remains correct at the centre of the scale only, which is why such ohmmeter designs always quote the accuracy "at centre scale only".

A more accurate type of ohmmeter has an electronic circuit that passes a constant current (I) through the resistance, and another circuit that measures the voltage (V) across the resistance. These measurements are then digitized with an analog digital converter (adc) after which a microcontroller or microprocessor make the division of the current and voltage according to Ohm's Law and then decode these to a display to offer the user a reading of the resistance value they're measuring at that instant. Since these type of meters already measure current,voltage and resistance all at once, these type of circuits are often used in digital multimeters.

Precision Ohmmeters

For high-precision measurements of very small resistances, the above types of meter are inadequate. This is partly because the change in deflection itself is small when the resistance measured is too small in proportion to the intrinsic resistance of the ohmmeter (which can be dealt with through current division), but mostly because the meter's reading is the sum of the resistance of the measuring leads, the contact resistances and the resistance being measured. To reduce this effect, a precision ohmmeter has four terminals, called Kelvin contacts. Two terminals carry the current from and to the meter, while the other two allow the meter to measure the voltage across the resistor. In this arrangement, the power source is connected in series with the resistance to be measured through the external pair of terminals, while the second pair connects in parallel with the galvanometer which measures the voltage drop. With this type of meter, any voltage drop due to the resistance of the first pair of leads and their contact resistances is ignored by the meter. This four terminal measurement technique is called Kelvin sensing, after William Thomson, Lord Kelvin, who invented the Kelvin bridge in 1861 to measure very low resistances. The Four-terminal sensing method can also be utilized to conduct accurate measurements of low resistances.

Ammeter

An ammeter (from Ampere Meter) is a measuring instrument used to measure the current in a circuit. Electric currents are measured in amperes (A), hence the name. Instruments used to measure smaller currents, in the milliampere or microampere range, are designated as *milliammeters* or *microammeters*. Early ammeters were laboratory instruments which relied on the Earth's magnetic field for operation. By the late 19th century, improved instruments were designed which could be mounted in any position and allowed accurate measurements in electric power systems. It is generally represented by letter 'A' in a circuit.

The relation between electric current, magnetic fields and physical forces was first noted by Hans Christian Ørsted who, in 1820, observed a compass needle was deflected from pointing North when a current flowed in an adjacent wire. The tangent galvanometer was used to measure currents using this effect, where the restoring force returning the pointer to the zero position was provided by the Earth's magnetic field. This made these instruments usable only when aligned with the Earth's field. Sensitivity of the instrument was increased by using additional turns of wire to multiply the effect – the instruments were called "multipliers".

The word *rheoscope* as a detector of electrical currents was coined by Sir Charles Wheatstone about 1840 but is no longer used to describe electrical instruments. The word makeup is similar to that of *rheostat* (also coined by Wheatstone) which was a device used to adjust the current in a circuit. Rheostat is a historical term for a variable resistance, though unlike rheoscope may still be encountered.

Types

Moving-coil

The D'Arsonval galvanometer is a moving coil ammeter. It uses magnetic deflection, where current passing through a coil placed in the magnetic field of a permanent magnet causes the coil to move. The modern form of this instrument was developed by Edward Weston, and uses two spiral springs to provide the restoring force. The uniform air gap between the iron core and the permanent magnet poles make the deflection of the meter linearly proportional to current. These meters have linear scales. Basic meter movements can have full-scale deflection for currents from about 25 microamperes to 10 milliamperes.

Because the magnetic field is polarised, the meter needle acts in opposite directions for each direction of current. A DC ammeter is thus sensitive to which way round it is connected; most are marked with a positive terminal, but some have centre-zero mechanisms and can display currents in either direction. A moving coil meter indicates the average (mean) of a varying current through it, which is zero for AC. For this reason moving-coil meters are only usable directly for DC, not AC.

This type of meter movement is extremely common for both ammeters and other meters derived from them, such as voltmeters and ohmmeters.

Moving Magnet

Moving magnet ammeters operate on essentially the same principle as moving coil, except that the coil is mounted in the meter case, and a permanent magnet moves the needle. Moving magnet Ammeters are able to carry larger currents than moving coil instruments, often several tens of Amperes, because the coil can be made of thicker wire and the current does not have to be carried by the hairsprings. Indeed, some Ammeters of this type do not have hairsprings at all, instead using a fixed permanent magnet to provide the restoring force.

Electrodynamic

An electrodynamic ammeter uses an electromagnet instead of the permanent magnet of the d'Arsonval movement. This instrument can respond to both alternating and direct current and also indicates true RMS for AC.

Moving-iron

Moving iron ammeters use a piece of iron which moves when acted upon by the electromagnetic force of a fixed coil of wire. The moving-iron meter was invented by Austrian engineer Friedrich Drexler in 1884. This type of meter responds to both direct and alternating currents (as opposed to the moving-coil ammeter, which works on direct current only). The iron element consists of a moving vane attached to a pointer, and a fixed vane, surrounded by a coil. As alternating or direct current flows through the coil and induces a magnetic field in both vanes, the vanes repel each other and the moving vane deflects against the restoring force provided by fine helical springs. The deflection of a moving iron meter is proportional to the square of the current. Consequently, such meters would normally have a non linear scale, but the iron parts are usually modified in shape to make the scale fairly linear over most of its range. Moving iron instruments indicate the RMS value of any AC waveform applied. Moving iron ammeters are commonly used to measure current in industrial frequency AC circuits.

Hot-wire

In a hot-wire ammeter, a current passes through a wire which expands as it heats. Although these instruments have slow response time and low accuracy, they were sometimes used in measuring radio-frequency current. These also measure true RMS for an applied AC.

Digital

In much the same way as the analogue ammeter formed the basis for a wide variety of derived meters, including voltmeters, the basic mechanism for a digital meter is a digital voltmeter mechanism, and other types of meter are built around this.

Digital ammeter designs use a shunt resistor to produce a calibrated voltage proportional to the current flowing. This voltage is then measured by a digital voltmeter, through use of an analog to digital converter (ADC); the digital display is calibrated to display the current through the shunt. Such instruments are often calibrated to indicate the RMS value for a sine wave only, but many designs will indicate true RMS within limitations of the wave crest factor.

Integrating

An integrating current meter calibrated in Ampere-hours or charge.

There is also a range of devices referred to as integrating ammeters. In these ammeters the current is summed over time, giving as a result the product of current and time; which is proportional to the electrical charge transferred with that current. These can be used for metering energy (the charge needs to be multiplied by the voltage to give energy) or for estimating the charge of a battery or capacitor.

Picoammeter

A picoammeter, or pico ammeter, measures very low electric current, usually from the picoampere range at the lower end to the milliampere range at the upper end. Picoammeters are used for sensitive measurements where the current being measured is below the theoretical limits of sensitivity of other devices, such as Multimeters.

Most picoammeters use a "virtual short" technique and have several different measurement ranges that must be switched between to cover multiple decades of measurement. Other modern picoammeters use log compression and a "current sink" method that eliminates range switching and associated voltage spikes. Special design and usage considerations must be observed in order to reduce leakage current which may swamp measurements such as special insulators and driven shields. Triaxial cable is often used for probe connections.

Application

The majority of ammeters are either connected in series with the circuit carrying the current to be measured (for small fractional amperes), or have their shunt resistors connected similarly in series. In either case, the current passes through the meter or (mostly) through its shunt. Ammeters must not be connected directly across a voltage source since their internal resistance is very low and excess current would flow. Ammeters are designed for a low voltage drop across their terminals, much less than one volt; the extra circuit losses produced by the ammeter are called its "burden" on the measured circuit.

Ordinary Weston-type meter movements can measure only milliamperes at most, because the springs and practical coils can carry only limited currents. To measure larger currents, a resistor called a *shunt* is placed in parallel with the meter. The resistances of shunts is in the integer to fractional milliohm range. Nearly all of the current flows through the shunt, and only a small fraction flows through the meter. This allows the meter to measure large currents. Traditionally, the meter used with a shunt has a full-scale deflection (FSD) of 50 mV, so shunts are typically designed to produce a voltage drop of 50 mV when carrying their full rated current.

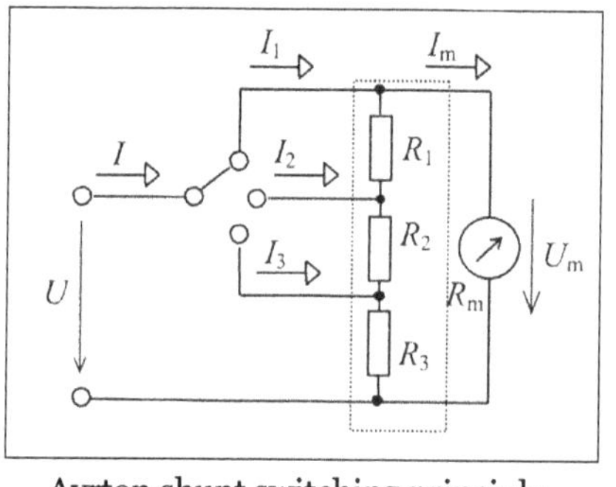

Ayrton shunt switching principle.

To make a multi-range ammeter, a selector switch can be used to connect one of a number of shunts across the meter. It must be a make-before-break switch to avoid damaging current surges through the meter movement when switching ranges.

A better arrangement is the Ayrton shunt or universal shunt, invented by William E. Ayrton, which does not require a make-before-break switch. It also avoids any inaccuracy because of contact resistance. In the figure, assuming for example, a movement with a full-scale voltage of 50 mV and desired current ranges of 10 mA, 100 mA, and 1 A, the resistance values would be: R1=4.5 ohms, R2=0.45 ohm, R3=0.05 ohm. And if the movement resistance is 1000 ohms, for example, R1 must be adjusted to 4.525 ohms.

Switched shunts are rarely used for currents above 10 amperes.

Zero-center ammeters are used for applications requiring current to be measured with both polarities, common in scientific and industrial equipment. Zero-center ammeters are also commonly placed in series with a battery. In this application, the charging of the battery deflects the needle to one side of the scale (commonly, the right side) and the discharging of the battery deflects the needle to the other side. A special type of zero-center ammeter for testing high currents in cars and trucks has a pivoted bar magnet that moves the pointer, and a fixed bar magnet to keep the pointer centered with no current. The magnetic field around the wire carrying current to be measured deflects the moving magnet.

Since the ammeter shunt has a very low resistance, mistakenly wiring the ammeter in parallel with a voltage source will cause a short circuit, at best blowing a fuse, possibly damaging the instrument and wiring, and exposing an observer to injury.

In AC circuits, a current transformer converts the magnetic field around a conductor into a small AC current, typically either 1 A or 5 A at full rated current, that can be easily read by a meter. In a similar way, accurate AC/DC non-contact ammeters have been constructed using Hall effect magnetic field sensors. A portable hand-held clamp-on ammeter is a common tool for maintenance of industrial and commercial electrical equipment, which is temporarily clipped over a wire to measure current. Some recent types have a parallel pair of magnetically soft probes that are placed on either side of the conductor.

Current Clamp

In electrical and electronic engineering, a current clamp or current probe is an electrical device with jaws which open to allow clamping around an electrical conductor. This allows measurement of the current in a conductor without the need to make physical contact with it, or to disconnect it for insertion through the probe. Current clamps are typically used to read the magnitude of alternating current (AC) and, with additional instrumentation, the phase and waveform can also be measured. Some clamp meters can measure currents of 1000 A and more. Hall effect and vane type clamps can also measure direct current (DC).

Types of Current Clamp

Current Transformer

A common form of current clamp comprises a split ring made of ferrite or soft iron. A wire coil is wound round one or both halves, forming one winding of a current transformer. The conductor it is clamped around forms the other winding. Like any transformer this type works only with AC or pulse waveforms, with some examples extending into the megahertz range.

When measuring current, the subject conductor forms the primary winding and the coil forms the secondary.

This type may also be used in reverse, to inject current into the conductor, for example in electromagnetic compatibility susceptibility testing to induce an interference current. Usually, the injection probe is specifically designed for this purpose. In this mode, the coil forms the primary and the test conductor the secondary.

Iron Vane

In the iron vane type, the magnetic flux in the core directly affects a moving iron vane, allowing both AC and DC to be measured, and gives a true root mean square (RMS) value for non-sinusoidal AC waveforms. Due to its physical size it is generally limited to power transmission frequencies up to around 100 Hz.

The vane is usually fixed directly to the display mechanism of an analogue (moving pointer) clamp meter. The calibration of the instrument is clearly non-linear.

Hall Effect

The Hall effect type is more sensitive and is able to measure both DC and AC, in some examples up to the kilohertz (thousands of hertz) range. This type was often used with oscilloscopes, and with high-end computerized digital multimeters, however, they are becoming common place for more general use.

Rogowski Coil

Resembling a current clamp in appearance and function is the Rogowski coil current sensor. This coreless transformer is used in clamp meters and power monitoring loggers. It has the advantage of better linearity, having no core to saturate, it can be made flexible, and does not require any magnetic or electrical contact at the opening end. The Rogowski coil gives a voltage proportional to the rate of change of current in the primary cable, so more signal processing is needed before the sensed values can be displayed.

Clamp Meter

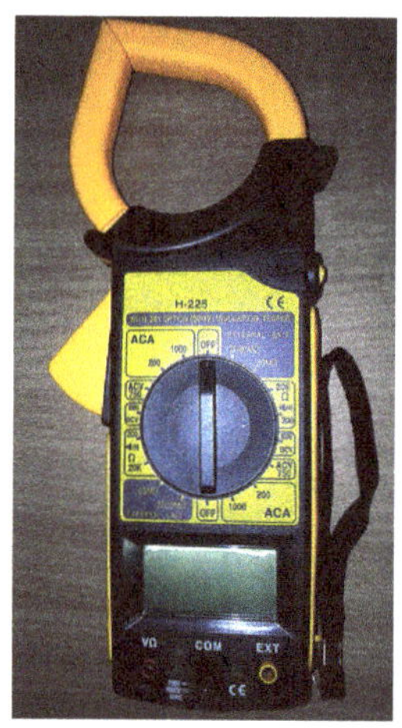

A digital clamp meter.

A multimeter with built-in clamp. Pushing the large button at the bottom opens the lower jaw of the clamp, allowing the clamp to be placed around a conductor.

An electrical meter with integral AC current clamp is known as a clamp meter, clamp-on ammeter, tong tester, or colloquially as an amp clamp.

A clamp meter measures the vector sum of the currents flowing in all the conductors passing through the probe, which depends on the phase relationship of the currents. Only one conductor is normally passed through the probe. In particular if the clamp is closed around a two-conductor cable carrying power to equipment, the same current flows down one conductor and up the other; the meter correctly reads a net current of zero. As electrical cables for equipment have both insulated conductors (and possibly an earth wire) bonded together, clamp meters are often used with what is essentially a short extension cord with the two conductors separated, so that the clamp can be placed around only one conductor of this extension.

A relatively recent development is a multi-conductor clamp meter with several sensor coils around the jaws of the clamp. This could be clamped around standard two- or three-conductor single-phase cables to provide a readout of the current flowing through the load, with no need to separate the conductors.

The reading produced by a conductor carrying a very low current can be increased by winding the conductor around the clamp several times; the meter reading divided by the number of turns is the current, with some loss of accuracy due to inductive effects.

Clamp meters are used by electricians, sometimes with the clamp incorporated into a general purpose multimeter.

It is simple to measure very high currents (hundreds of amperes) with the appropriate current transformer. Accurate measurement of low currents (a few milliamperes) with a current transformer clamp is more difficult. The range of any given meter can be extended by passing the conductor through the jaw multiple times. For example a 0–200 A meter can be turned into a 0–20 A meter by winding the conductor 10 times around the jaw's core.

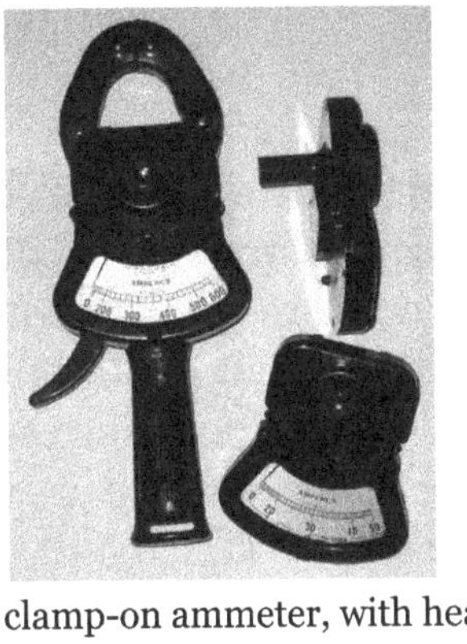

Columbia iron vane type clamp-on ammeter, with heads for different ranges.

Less-expensive clamp meters use a rectifier circuit which actually reads mean current, but is calibrated to display the RMS current corresponding to the measured mean, giving a correct RMS reading only if the current is a sine wave. For other waveforms readings will be incorrect; when these simpler meters are used with non-sinusoidal loads such as the ballasts used with fluorescent lamps or high-intensity discharge lamps or most modern computer and electronic equipment, readings can be quite inaccurate. Meters which respond to true RMS rather than mean current are described as "true RMS".

Typical hand-held Hall effect units can read currents as low as 200 mA, and units that can read down to 1 mA are available.

The Columbia tong test ammeter (illustrated) is an example of the iron vane type, used for measuring large AC currents up to 1000 amperes. The iron jaws of the meter direct the magnetic field surrounding the conductor to an iron vane that is attached to the needle of the meter. The iron vane moves in proportion to the strength of the magnetic field, and thus produces a meter indication proportional to the current. This type of ammeter can measure both AC and DC currents and provides a true RMS current measurement of non-sinusoidal or distorted AC waveforms. Interchangeable meter movements can be installed in the clamping assembly to provide various full-scale current values up to 1000 amperes. The iron vane is in a small cylinder that is inserted in a space at the hinged end of the clamp-on jaws. Several jaw sizes are available for clamping around large conductors and bus bars up to 4 $\frac{1}{2}$ inches (110 mm) wide. As the illustration shows, the scale is very non-linear and unsuitable for measuring low currents, with currents of less than half the full-scale deflection crammed into a short section of the dial.

Power Meter and Energy Analyzer

Clamp probes are used with some meters to measure electrical power and energy. The clamp measures the current and other circuitry the voltage; the true power is the product of the instantaneous voltage and current integrated over a cycle. Comprehensive meters designed to measure many parameters of electrical energy (power factor, distortion, instantaneous power as a function of time, phase relationships, etc.), energy analyzers, use this principle. A single clamp is used for single-phase measurements; with an appropriate instrument with three clamps, measurements may be made on three-phase power systems.

Multimeter

A multimeter or a multitester, also known as a VOM (volt-ohm-milliammeter), is an electronic measuring instrument that combines several measurement functions in one unit. A typical multimeter can measure voltage, current, and resistance. Analog multimeters uses a microammeter with a moving pointer to display readings. Digital multimeters (DMM, DVOM) have a numeric display, and may also show a graphical bar representing the measured value. Digital multimeters are now far more common due to their lower cost and greater precision, but analog multimeters are still preferable in some cases, for example when monitoring a rapidly varying value.

A multimeter can be a hand-held device useful for basic fault finding and field service work, or a bench instrument which can measure to a very high degree of accuracy. Multimeters are available in a wide range of features and prices.

1920s pocket multimeter.

Avometer Model 8.

The first moving-pointer current-detecting device was the galvanometer in 1820. These were used to measure resistance and voltage by using a Wheatstone bridge, and comparing the unknown quantity to a reference voltage or resistance. While useful in the lab, the devices were very slow and impractical in the field. These galvanometers were bulky and delicate.

The D'Arsonval–Weston meter movement uses a moving coil which carries a pointer and rotates on pivots or a taut band ligament. The coil rotates in a permanent magnetic field and is restrained by fine spiral springs which also serve to carry current into the moving coil. It gives proportional measurement rather than just detection, and deflection is independent of the orientation of the meter. Instead of balancing a bridge, values could be directly read off the instrument's scale, which made measurement quick and easy.

The basic moving coil meter is suitable only for direct current measurements, usually in the range of 10 μA to 100 mA. It is easily adapted to read heavier currents by using shunts (resistances in parallel with the basic movement) or to read voltage using series resistances known as multipliers. To read alternating currents or voltages, a rectifier is needed. One of the earliest suitable rectifiers was the copper oxide rectifier developed and manufactured by Union Switch & Signal Company, Swissvale, Pennsylvania, later part of Westinghouse Brake and Signal Company, from 1927.

Multimeters were invented in the early 1920s as radio receivers and other vacuum tube electronic devices became more common. The invention of the first multimeter is attributed to British Post Office engineer, Donald Macadie, who became dissatisfied with the need to carry many separate instruments required for maintenance of telecommunications circuits. Macadie invented an instrument which could measure amperes (amps), volts and ohms, so the multifunctional meter was then named Avometer. The meter comprised a moving coil meter, voltage and precision resistors, and switches and sockets to select the range.

The Automatic Coil Winder and Electrical Equipment Company (ACWEECO), founded in 1923, was set up to manufacture the Avometer and a coil winding machine also designed and patented by MacAdie. Although a shareholder of ACWEECO, Mr MacAdie continued to work for the Post Office until his retirement in 1933. His son, Hugh S. MacAdie, joined ACWEECO in 1927 and became Technical Director. The first AVO was put on sale in 1923, and many of its features remained almost unaltered through to the last Model 8.

General Properties of Multimeters

Any meter will load the circuit under test to some extent. For example, a multimeter using a moving coil movement with full-scale deflection current of 50 microamps (μA), the highest sensitivity commonly available, must draw at least 50 μA from the circuit under test for the meter to reach the top end of its scale. This may load a high-impedance circuit so much as to affect the circuit, thereby giving a low reading. The full-scale deflection current may also be expressed in terms of "ohms per volt" (Ω/V). The ohms per volt figure is often called the "sensitivity" of the instrument. Thus a meter with a 50 μA movement will have a "sensitivity" of 20,000 Ω/V. "Per volt" refers to the fact that the impedance the meter presents to the circuit under test will be 20,000 Ω multiplied by the full-scale voltage to which the meter is set. For example, if the meter is set to a range of 300 V full scale, the meter's impedance will be 6 MΩ. 20,000 Ω/V is the best (highest) sensitivity available for typical analog multimeters that lack internal amplifiers. For meters that do have internal amplifiers (VTVMs, FETVMs, etc.), the input impedance is fixed by the amplifier circuit.

The first Avometer had a sensitivity of 60 Ω/V, three direct current ranges (12 mA, 1.2 A, and 12 A), three direct voltage ranges (12, 120, and 600 V or optionally 1,200 V), and a 10,000 Ω resistance range. An improved version of 1927 increased this to 13 ranges and 166.6 Ω/V (6 mA) movement. A "Universal" version having additional alternating current and alternating voltage ranges was offered from 1933 and in 1936 the dual-sensitivity Avometer Model 7 offered 500 and 100 Ω/V. Between the mid 1930s until the 1950s, 1,000 Ω/V became a de facto standard of sensitivity for radio work and this figure was often quoted on service sheets. However, some manufacturers such as Simpson, Triplett and Weston, all in the USA, produced 20,000 Ω/V VOMs before the Second World War and some of these were exported. After 1945–46, 20,000 Ω/V became the expected standard for electronics, but some makers offered even more sensitive instruments. For industrial and other "heavy-current" use low sensitivity multimeters continued to be produced and these were considered more robust than the more sensitive types.

High-quality analog (analogue) multimeters continue to be made by several manufacturers, including Chauvin Arnoux (France), Gossen Metrawatt (Germany), and Simpson and Triplett (USA).

Pocket-watch-style meters were in widespread use in the 1920s. The metal case was typically connected to the negative connection, an arrangement that caused numerous electric shocks. The technical specifications of these devices were often crude, for example the one illustrated has a resistance of just 33 Ω/V, a non-linear scale and no zero adjustment.

Vacuum tube voltmeters or valve voltmeters (VTVM, VVM) were used for voltage measurements in electronic circuits where high input impedance was necessary. The VTVM had a fixed input impedance of typically 1 MΩ or more, usually through use of a cathode follower input circuit, and thus did not significantly load the circuit being tested. VTVMs were used before the introduction of electronic high-impedance analog transistor and field effect transistor voltmeters (FETVOMs). Modern digital meters (DVMs) and some modern analog meters also use electronic input circuitry to achieve high input impedance—their voltage ranges are functionally equivalent to VTVMs. The input impedance of some poorly designed DVMs (especially some early designs) would vary over the course of a sample-and-hold internal measurement cycle, causing disturbances to some sensitive circuits under test.

Additional scales such as decibels, and measurement functions such as capacitance, transistor gain, frequency, duty cycle, display hold, and continuity which sounds a buzzer when the measured resistance is small have been included on many multimeters. While multimeters may be supplemented by more specialized equipment in a technician's toolkit, some multimeters include additional functions for specialized applications (temperature with a thermocouple probe, inductance, connectivity to a computer, speaking measured value, etc.).

Operation

A 4 $\frac{1}{2}$-digit digital multimeter, the Fluke 87V.

A multimeter is the combination of a DC voltmeter, AC voltmeter, ammeter, and ohmmeter. An un-amplified analog multimeter combines a meter movement, range resistors and switches; VTVMs are amplified analog meters and contain active circuitry.

For an analog meter movement, DC voltage is measured with a series resistor connected between the meter movement and the circuit under test. A switch (usually rotary) allows greater resistance to be inserted in series with the meter movement to read higher voltages. The product of the basic full-scale deflection current of the movement, and the sum of the series resistance and the movement's own resistance, gives the full-scale voltage of the range. As an example, a meter movement that required 1 mA for full-scale deflection, with an internal resistance of 500 Ω, would, on a 10 V range of the multimeter, have 9,500 Ω of series resistance.

For analog current ranges, matched low-resistance shunts are connected in parallel with the meter movement to divert most of the current around the coil. Again for the case of a hypothetical 1 mA, 500 Ω movement on a 1 A range, the shunt resistance would be just over 0.5 Ω.

Moving coil instruments can respond only to the average value of the current through them. To measure alternating current, which changes up and down repeatedly, a rectifier is inserted in the circuit so that each negative half cycle is inverted; the result is a varying and nonzero DC voltage whose maximum value will be half the AC peak to peak voltage, assuming a symmetrical waveform. Since the rectified average value and the root mean square (RMS) value of a waveform are only the same for a square wave, simple rectifier-type circuits can only be calibrated for sinusoidal waveforms. Other wave shapes require a different calibration factor to relate RMS and average value. This type of circuit usually has fairly limited frequency range. Since practical rectifiers have non-zero voltage drop, accuracy and sensitivity is poor at low AC voltage values.

To measure resistance, switches arrange for a small battery within the instrument to pass a current through the device under test and the meter coil. Since the current available depends on the state of charge of the battery which changes over time, a multimeter usually has an adjustment for the ohm scale to zero it. In the usual circuits found in analog multimeters, the meter deflection is inversely proportional to the resistance, so full-scale will be 0 Ω, and higher resistance will correspond to smaller deflections. The ohms scale is compressed, so resolution is better at lower resistance values.

Amplified instruments simplify the design of the series and shunt resistor networks. The internal resistance of the coil is decoupled from the selection of the series and shunt range resistors; the series network thus becomes a voltage divider. Where AC measurements are required, the rectifier can be placed after the amplifier stage, improving precision at low range.

Digital instruments, which necessarily incorporate amplifiers, use the same principles as analog instruments for resistance readings. For resistance measurements, usually a small constant current is passed through the device under test and the digital multimeter reads the resultant voltage drop; this eliminates the scale compression found in analog meters, but requires a source of precise current. An autoranging digital multimeter can automatically adjust the scaling network so the measurement circuits use the full precision of the A/D converter.

In all types of multimeters, the quality of the switching elements is critical to stable and accurate measurements. The best DMMs use gold plated contacts in their switches; less expensive meters use nickel plating or none at all, relying on printed circuit board solder traces for the contacts. Accuracy and stability (e.g., temperature variation, or aging, or voltage/current history) of a meter's internal resistors (and other components) is a limiting factor in long-term accuracy and precision of the instrument.

Measured Values

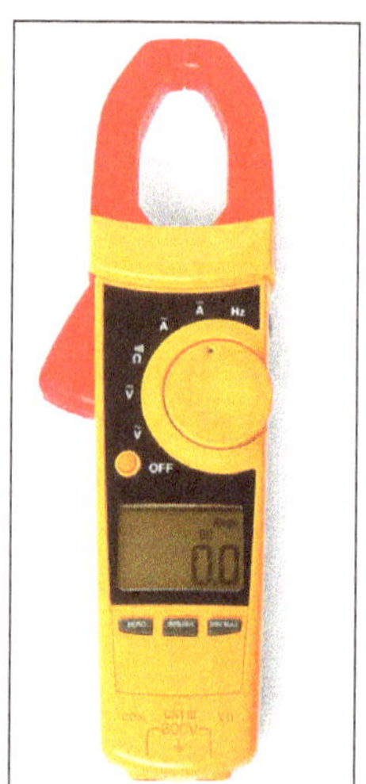

A clamp meter.

Contemporary multimeters can measure many values. The most common are:

- Voltage, alternating and direct, in volts.

- Current, alternating and direct, in amperes.

  The frequency range for which AC measurements are accurate is important, depends on the circuitry design and construction, and should be specified, so users can evaluate the readings they take. Some meters measure currents as low as milliamps or even microamps. All meters have a burden voltage (caused by the combination of the shunt used and the meter's circuit design), and some (even expensive ones) have sufficiently high burden voltages that low current readings are seriously impaired. Meter specifications should include the burden voltage of the meter.

- Resistance in ohms.

Additionally, some multimeters also measure:

- Capacitance in farads, but usually the limitations of the range are between a few hundred or thousand micro farads and a few pico farads. Very few general purpose multimeters can measure other important aspects of capacitor status such as ESR, dissipation factor, or leakage.
- Conductance in siemens, which is the inverse of the resistance measured.
- Decibels in circuitry, rarely in sound.
- Duty cycle as a percentage.
- Frequency in hertz.
- Inductance in henries. Like capacitance measurement, this is usually better handled by a purpose designed inductance / capacitance meter.
- Temperature in degrees Celsius or Fahrenheit, with an appropriate temperature test probe, often a thermocouple.

Digital multimeters may also include circuits for:

- Continuity tester; a buzzer sounds when a circuit's resistance is low enough (just how low is enough varies from meter to meter), so the test must be treated as inexact.
- Diodes (measuring forward drop of diode junctions).
- Transistors (measuring current gain and other parameters in some kinds of transistors).
- Battery checking for simple 1.5 V and 9 V batteries. This is a current-loaded measurement, which simulates in-use battery loads; normal voltage ranges draw very little current from the battery.

Various sensors can be attached to (or included in) multimeters to take measurements such as:

- Light level.
- Sound pressure level.
- Acidity/alkalinity(ph).

- Relative humidity.
- Very small current flow (down to nanoamps with some adapters).
- Very small resistances (down to micro ohms for some adapters).
- Large currents — adapters are available which use inductance (ac current only) or hall effect sensors (both ac and dc current), usually through insulated clamp jaws to avoid direct contact with high current capacity circuits which can be dangerous, to the meter and to the operator.
- Very high voltages — adapters are available which form a voltage divider with the meter's internal resistance, allowing measurement into the thousands of volts. However, very high voltages often have surprising behavior, aside from effects on the operator (perhaps fatal); high voltages which actually reach a meter's internal circuitry may internal damage parts, perhaps destroying the meter or permanently ruining its performance.

Resolution

Resolution and Accuracy

The resolution of a multimeter is the smallest part of the scale which can be shown, which is scale dependent. On some digital multimeters it can be configured, with higher resolution measurements taking longer to complete. For example, a multimeter that has a 1 mV resolution on a 10 V scale can show changes in measurements in 1 mV increments.

Absolute accuracy is the error of the measurement compared to a perfect measurement. Relative accuracy is the error of the measurement compared to the device used to calibrate the multimeter. Most multimeter datasheets provide relative accuracy. To compute the absolute accuracy from the relative accuracy of a multimeter add the absolute accuracy of the device used to calibrate the multimeter to the relative accuracy of the multimeter.

Digital

The resolution of a multimeter is often specified in the number of decimal digits resolved and displayed. If the most significant digit cannot take all values from 0 to 9 it is generally, and confusingly, termed a fractional digit. For example, a multimeter which can read up to 19999 (plus an embedded decimal point) is said to read 4 ½ digits.

By convention, if the most significant digit can be either 0 or 1, it is termed a half-digit; if it can take higher values without reaching 9 (often 3 or 5), it may be called three-quarters of a digit. A 5 ½-digit multimeter would display one "half digit" that could only display 0 or 1, followed by five digits taking all values from 0 to 9. Such a meter could show positive or negative values from 0 to 199999. A 3 ¾-digit meter can display a quantity from 0 to 3999 or 5999, depending on the manufacturer.

While a digital display can easily be extended in resolution, the extra digits are of no value if not accompanied by care in the design and calibration of the analog portions of the multimeter. Meaningful (i.e., high-accuracy) measurements require a good understanding of the instrument specifications, good control of the measurement conditions, and traceability of the calibration of the instrument. However, even if its resolution exceeds the accuracy, a meter can be useful for comparing measurements. For example, a meter reading 5 ½ stable digits may indicate that one nominally 100 kΩ resistor is about 7 Ω greater than another, although the error of each measurement is 0.2% of reading plus 0.05% of full-scale value.

Specifying "display counts" is another way to specify the resolution. Display counts give the largest number, or the largest number plus one (to include the display of all zeros) the multimeter's display can show, ignoring the decimal separator. For example, a 5 ½-digit multimeter can also be specified as a 199999 display count or 200000 display count multimeter. Often the display count is just called the 'count' in multimeter specifications.

The accuracy of a digital multimeter may be stated in a two-term form, such as "±1% of reading +2 counts", reflecting the different sources of error in the instrument.

Analog

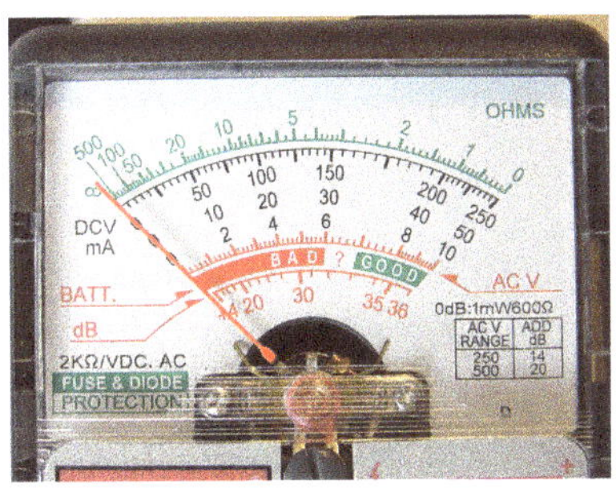

Display face of an analog multimeter.

Analog meters are older designs, but are still preferred by many engineers and troubleshooters. One reason for this is that analog meters are more sensitive to changes in the circuit that is being measured. A digital multimeter samples the quantity being measured over time, and then displays it. Analog multimeters continuously read the test value. If there are slight changes in readings, the needle of an analog multimeter will track them, while digital multimeters may miss them, or be difficult to read. This continuous tracking feature becomes important when testing capacitors or coils, for example. A properly functioning capacitor should allow current to flow when voltage is applied, then the current slowly decreases to zero and this "signature" is easy to see on an analog multimeter but not on a digital multimeter. This is similar when testing a coil, except the current starts low and increases.

Resistance measurements on an analog meter, in particular, can be of low precision due to the typical resistance measurement circuit which compresses the scale heavily at the higher resistance values. Inexpensive analog meters may have only a single resistance scale, seriously restricting the range of precise measurements. Typically, an analog meter will have a panel adjustment to set the zero-ohms calibration of the meter, to compensate for the varying voltage of the meter battery, and the resistance of the meter's test leads.

Accuracy

Digital multimeters generally take measurements with accuracy superior to their analog counterparts. Standard analog multimeters measure with typically ±3% accuracy, though instruments of higher accuracy are made. Standard portable digital multimeters are specified to have an accuracy of typically ±0.5% on the DC voltage ranges. Mainstream bench-top multimeters are available with specified accuracy of better than ±0.01%. Laboratory grade instruments can have accuracies of a few parts per million.

Accuracy figures need to be interpreted with care. The accuracy of an analog instrument usually refers to full-scale deflection; a measurement of 30 V on the 100 V scale of a 3% meter is subject to an error of 3 V, 10% of the reading. Digital meters usually specify accuracy as a percentage of reading plus a percentage of full-scale value, sometimes expressed in counts rather than percentage terms.

Quoted accuracy is specified as being that of the lower millivolt (mV) DC range, and is known as the "basic DC volts accuracy" figure. Higher DC voltage ranges, current, resistance, AC and other ranges will usually have a lower accuracy than the basic DC volts figure. AC measurements only meet specified accuracy within a specified range of frequencies.

Manufacturers can provide calibration services so that new meters may be purchased with a certificate of calibration indicating the meter has been adjusted to standards traceable to, for example, the US National Institute of Standards and Technology (NIST), or other national standards organization.

Test equipment tends to drift out of calibration over time, and the specified accuracy cannot be relied upon indefinitely. For more expensive equipment, manufacturers and third parties provide calibration services so that older equipment may be recalibrated and recertified. The cost of such services is disproportionate for inexpensive equipment; however extreme accuracy is not required for most routine testing. Multimeters used for critical measurements may be part of a metrology program to assure calibration.

A multimeter can be assumed to be "average responding" to AC waveforms unless stated as being a "true RMS" type. An average responding multimeter will only meet its specified accuracy on AC volts and amps for purely sinusoidal waveforms. A True RMS responding multimeter on the other hand will meet its specified accuracy on AC volts and current with any waveform type up to a specified crest factor; RMS performance is sometimes claimed for meters which report accurate RMS readings only at certain frequencies (usually low) and with certain waveforms (essentially always sine waves).

A meter's AC voltage and current accuracy may have different specifications at different frequencies.

Sensitivity and Input Impedance

When used for measuring voltage, the input impedance of the multimeter must be very high compared to the impedance of the circuit being measured; otherwise circuit operation may be changed, and the reading will also be inaccurate.

Meters with electronic amplifiers (all digital multimeters and some analog meters) have a fixed input impedance that is high enough not to disturb most circuits. This is often either one or ten megohms; the standardization of the input resistance allows the use of external high-resistance probes which form a voltage divider with the input resistance to extend voltage range up to tens of thousands of volts. High-end multimeters generally provide an input impedance greater than 10 GΩ for ranges less than or equal to 10 V. Some high-end multimeters provide >10 Gigaohms of impedance to ranges greater than 10 V.

Most analog multimeters of the moving-pointer type are unbuffered, and draw current from the circuit under test to deflect the meter pointer. The impedance of the meter varies depending on the basic sensitivity of the meter movement and the range which is selected. For example, a meter with a typical 20,000 Ω/V sensitivity will have an input resistance of 2 MΩ on the 100 V range (100 V × 20,000 Ω = 2,000,000 Ω). On every range, at full-scale voltage of the range, the full current required to deflect the meter movement is taken from the circuit under test. Lower sensitivity meter movements are acceptable for testing in circuits where source impedances are low compared to the meter impedance, for example, power circuits; these meters are more rugged mechanically. Some measurements in signal circuits require higher sensitivity movements so as not to load the circuit under test with the meter impedance.

Sensitivity should not be confused with resolution of a meter, which is defined as the lowest signal change (voltage, current, resistance and so on) that can change the observed reading.

For general-purpose digital multimeters, the lowest voltage range is typically several hundred millivolts AC or DC, but the lowest current range may be several hundred microamperes, although instruments with greater current sensitivity are available. Multimeters designed for (mains) "electrical" use instead of general electronics engineering use will typically forego the microamps current ranges.

Measurement of low resistance requires lead resistance (measured by touching the test probes together) to be subtracted for best accuracy. This can be done with the "delta", "zero", or "null" feature of many digital multimeters. Contact pressure to the device under test and cleanliness of the surfaces can affect measurements of very low resistances. Some meters offer a four wire test where two probes supply the source voltage and the others take measurement. Using a very high impedance allows for very low voltage drop in the probes and resistance of the source probes is ignored resulting in very accurate results.

The upper end of multimeter measurement ranges varies considerably; measurements over perhaps 600 volts, 10 amperes, or 100 megohms may require a specialized test instrument.

Burden Voltage

Every inline series-connected ammeter, including a multimeter in a current range, has a certain resistance. Most multimeters inherently measure voltage, and pass a current to be measured through a shunt resistance, measuring the voltage developed across it. The voltage drop is known as the burden voltage, specified in volts per ampere. The value can change depending on the range the meter sets, since different ranges usually use different shunt resistors.

The burden voltage can be significant in very low-voltage circuit areas. To check for its effect on accuracy and on external circuit operation the meter can be switched to different ranges; the current reading should be the same and circuit operation should not be affected if burden voltage is not a problem. If this voltage is significant it can be reduced (also reducing the inherent accuracy and precision of the measurement) by using a higher current range.

Alternating Current Sensing

Since the basic indicator system in either an analog or digital meter responds to DC only, a multimeter includes an AC to DC conversion circuit for making alternating current measurements. Basic meters utilize a rectifier circuit to measure the average or peak absolute value of the voltage, but are calibrated to show the calculated root mean square (RMS) value for a sinusoidal waveform; this will give correct readings for alternating current as used in power distribution. User guides for some such meters give correction factors for some simple non-sinusoidal waveforms, to allow the correct root mean square (RMS) equivalent value to be calculated. More expensive multimeters include an AC to DC converter that measures the true RMS value of the waveform within certain limits; the user manual for the meter may indicate the limits of the crest factor and frequency for which the meter calibration is valid. RMS sensing is necessary for measurements on non-sinusoidal periodic waveforms, such as found in audio signals and variable-frequency drives.

Digital Multimeters (DMM or DVOM)

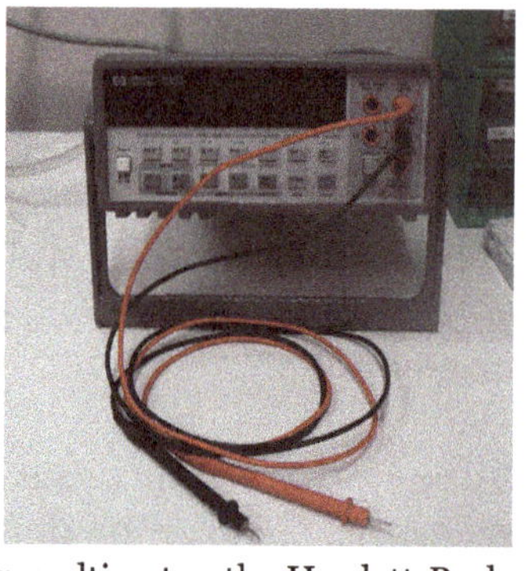

A bench-top multimeter, the Hewlett-Packard 34401a.

A digital multimeter (DMM) is a device that provides multiple types of measurements within one device. DMMs uses a more accurate and readable digital output than an analog solution. Without using a DMM to immediately measure live voltage and currents, there is a risk of painful shocks that can even be lethal. Therefore, it poses as a serious safety hazard.

Modern multimeters are often digital due to their accuracy, durability and extra features. In a digital multimeter the signal under test is converted to a voltage and an amplifier with electronically controlled gain preconditions the signal. A digital multimeter displays the quantity measured as a number, which eliminates parallax errors.

Modern digital multimeters may have an embedded computer, which provides a wealth of convenience features. Measurement enhancements available include:

- Auto-ranging, which selects the correct range for the quantity under test so that the most significant digits are shown. For example, a four-digit multimeter would automatically select an appropriate range to display 12.34 mV instead of 0.012 V, or overloading. Auto-ranging meters usually include a facility to hold the meter to a particular range, because a measurement that causes frequent range changes can be distracting to the user.
- Auto-polarity for direct-current readings, shows if the applied voltage is positive (agrees with meter lead labels) or negative (opposite polarity to meter leads).

- Sample and hold, which will latch the most recent reading for examination after the instrument is removed from the circuit under test.
- Current-limited tests for voltage drop across semi conductor junctions. While not a replacement for a proper transistor tester, and most certainly not for a swept curve tracer type, this facilitates testing diodes and a variety of transistor types.
- A graphic representation of the quantity under test, as a bar graph. This makes go/no-go testing easy, and also allows spotting of fast-moving trends.
- A low-bandwidth oscilloscope.
- Automotive circuit testers, including tests for automotive timing and dwell signals (dwell and engine rpm testing is usually available as an option and is not included in the basic automotive DMMs).
- Simple data acquisition features to record maximum and minimum readings over a given period, or to take a number of samples at fixed intervals.
- Integration with tweezers for surface-mount technology.
- A combined LCR meter for small-size SMD and through-hole components.

Modern meters may be interfaced with a personal computer by IrDA links, RS-232 connections, USB, or an instrument bus such as IEEE-488. The interface allows the computer to record measurements as they are made. Some DMMs can store measurements and upload them to a computer.

The first digital multimeter was manufactured in 1955 by Non Linear Systems. It is claimed that the first *handheld* digital multimeter was developed by Frank Bishop of Intron Electronics in 1977, which at the time presented a major breakthrough for servicing and fault finding in the field.

Analog Multimeters

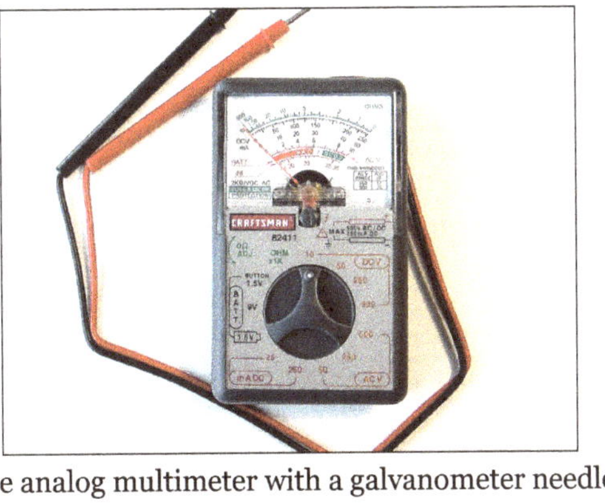

Inexpensive analog multimeter with a galvanometer needle display.

A multimeter may be implemented with a galvanometer meter movement, or less often with a bargraph or simulated pointer such as an LCD or vacuum fluorescent display. Analog multimeters are common; a quality analog instrument will cost about the same as a DMM. Analog multimeters have the precision and reading accuracy limitations described above, and so are not built to provide the same accuracy as digital instruments.

Analog meters are also useful where the trend of a measurement is more important than an exact value obtained at a particular moment. A change in angle or in a proportion is easier to interpret than a change in a digital readout. For this reason, digital multimeters often approximate this with a bargraph (typically with a more rapid reading response than the primary readout); the most effective of these are arranged in an arc, to simulate the pointer of an analog meter.

Analog meter movements are inherently more fragile physically and electrically than digital meters. Many analog multimeters feature a range switch position marked "off" to protect the meter movement during transportation which places placing a low resistance across the meter movement, resulting in dynamic braking. Meter movements as separate components may be protected in the same manner by connecting a shorting or jumper wire between the terminals when not in use. Meters which feature a shunt across the winding such as an ammeter may not require further resistance to arrest uncontrolled movements of the meter needle because of the low resistance of the shunt.

The meter movement in a moving pointer analog multimeter is practically always a moving-coil galvanometer of the d'Arsonval type, using either jeweled pivots or taut bands to support the moving coil. In a basic analog multimeter the current to deflect the coil and pointer is drawn from the circuit being measured; it is usually an advantage to minimize the current drawn from the circuit, which implies delicate mechanisms. The sensitivity of an analog multimeter is given in units of ohms per volt. For example, a very low-cost multimeter with a sensitivity of 1,000 Ω/V would draw 1 mA from a circuit at full-scale deflection. More expensive, (and mechanically more delicate) multimeters typically have sensitivities of 20,000 ohms per volt and sometimes higher, with 50,000 ohms per volt (drawing 20 microamperes at full scale) being about the upper limit for a portable, general purpose, non-amplified analog multimeter.

To avoid the loading of the measured circuit by the current drawn by the meter movement, some analog multimeters use an amplifier inserted between the measured circuit and the meter movement. While this increases the expense and complexity of the meter, by use of vacuum tubes or field effect transistors the input resistance can be made very high and independent of the current required to operate the meter movement coil. Such amplified multimeters are called VTVMs (vacuum tube voltmeters), TVMs (transistor volt meters), FET-VOMs, and similar names.

Because of the absence of amplification, ordinary analog multimeter are typically less susceptible to radio frequency interference, and so continue to have a prominent place in some fields even in a world of more accurate and flexible electronic multimeters.

Probes

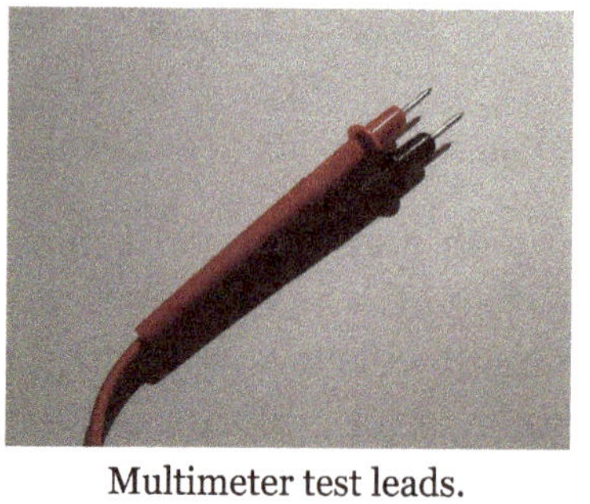

Multimeter test leads.

A multimeter can use many different test probes to connect to the circuit or device under test. Crocodile clips, retractable hook clips, and pointed probes are the three most common types. Tweezer probes are used for closely spaced test points, as for instance surface-mount devices. The connectors are attached to flexible, well insulated leads terminated with connectors appropriate for the meter. Probes are connected to portable meters typically by shrouded or recessed banana jacks, while benchtop meters may use banana jacks or BNC connectors. 2 mm plugs and binding posts have also been used at times, but are less commonly used today. Indeed, safety ratings now require shrouded banana jacks.

The banana jacks are typically placed with a standardized center-to-center distance of $\frac{3}{4}$ in (19 mm), to allow standard adapters or devices such as voltage multiplier or thermocouple probes to be plugged in.

Clamp meters clamp around a conductor carrying a current to measure without the need to connect the meter in series with the circuit, or make metallic contact at all. Those for AC measurement use the transformer principle; clamp-on meters to measure small current or direct current require more exotic sensors like for example hall effect based systems that measure the nonchanging magnetic field to determine the current.

Safety

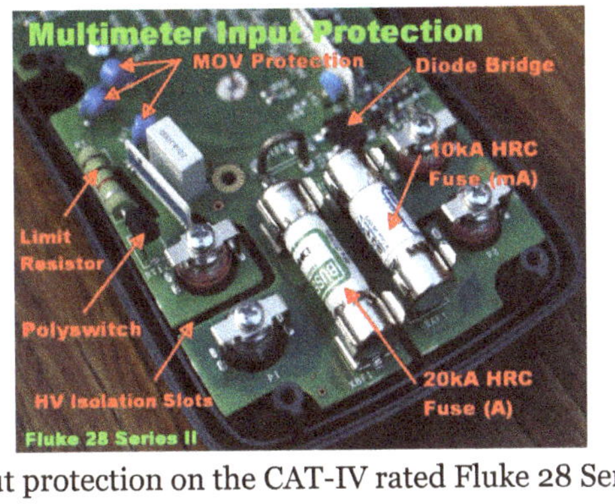

An example of input protection on the CAT-IV rated Fluke 28 Series II Multimeter.

Most multimeters include a fuse, or two fuses, which will sometimes prevent damage to the multimeter from a current overload on the highest current range. (For added safety, test leads with fuses built in are available.) A common error when operating a multimeter is to set the meter to measure resistance or current, and then connect it directly to a low-impedance voltage source. Unfused meters are often quickly destroyed by such errors; fused meters often survive. Fuses used in meters must carry the maximum measuring current of the instrument, but are intended to disconnect if operator error exposes the meter to a low-impedance fault. Meters with inadequate or unsafe fusing were not uncommon; this situation has led to the creation of the IEC61010 categories to rate the safety and robustness of meters.

Digital meters are rated into four categories based on their intended application, as set forth by IEC 61010-1 and echoed by country and regional standards groups such as the CEN EN61010 standard.

- Category I: used where equipment is not directly connected to the mains.
- Category II: used on single phase mains final subcircuits.

- Category III: used on permanently installed loads such as distribution panels, motors, and three-phase appliance outlets.
- Category IV: used on locations where fault current levels can be very high, such as supply service entrances, main panels, supply meters, and primary over-voltage protection equipment.

Each Category rating also specifies maximum safe transient voltages for selected measuring ranges in the meter. Category-rated meters also feature protections from over-current faults. On meters that allow interfacing with computers, optical isolation may be used to protect attached equipment against high voltage in the measured circuit.

Good quality multimeters designed to meet Category II and above standards include high rupture capacity (HRC) ceramic fuses typically rated at more than 20 kA capacity; these are much less likely to fail explosively than more common glass fuses. They will also include high energy overvoltage MOV (Metal Oxide Varistor) protection, and circuit over-current protection in the form of a Polyswitch.

It is safest (for both multimeter, circuit under test, and the operator) to disconnect a component from its circuit, and almost always, to remove power from the device being investigated. Removing all power connections from a piece of mains powered equipment before testing (and ensuring that all large capacitance devices are discharged safely) is the safest choice. Leaving equipment attached to the mains supply while making measurements should be only a very carefully considered alternative choice. Amongst other issues, there are interactions between the ground arrangements for wall powered test equipment, and the device under test, which are unsafe, and can damage test equipment and the device under test. This is especially so when there is a fault, suspected or not, in any of the interconnected devices. Battery powered test equipment may be the safest choice in such situations.

Meters intended for testing in hazardous locations or for use on blasting circuits may require use of a manufacturer-specified battery to maintain their safety rating.

DMM Alternatives

A quality general-purpose electronics DMM is generally considered adequate for measurements at signal levels greater than 1 mV or 1 μA, or below about 100 MΩ; these values are far from the theoretical limits of sensitivity, and are of considerable interest in some circuit design situations. Other instruments—essentially similar, but with higher sensitivity—are used for accurate measurements of very small or very large quantities. These include nanovoltmeters, electrometers (for very low currents, and voltages with very high source resistance, such as 1 TΩ) and picoammeters. Accessories for more typical multimeters permit some of these measurements, as well. Such measurements are limited by available technology, and ultimately by inherent thermal noise.

Power Supply

Analog meters can measure voltage and current by using power from the test circuit, but require a supplementary internal voltage source for resistance testing, while electronic meters always require an internal power supply to run their internal circuitry. Hand-held meters use batteries, while bench meters usually use mains power; either arrangement allows the meter to test devices. Testing often requires that the component under test be isolated from the circuit in which they are mounted, as otherwise stray or leakage current paths may distort measurements. In some cases, the voltage from the multimeter may turn active devices on, distorting a measurement, or in extreme cases even damage an element in the circuit being investigated.

Network Analyzer

A network analyzer is an instrument that measures the network parameters of electrical networks. Today, network analyzers commonly measure s–parameters because reflection and transmission of electrical networks are easy to measure at high frequencies, but there are other network parameter sets such as y-parameters, z-parameters, and h-parameters. Network analyzers are often used to characterize two-port networks such as amplifiers and filters, but they can be used on networks with an arbitrary number of ports.

Network analyzers are used mostly at high frequencies; operating frequencies can range from 5 Hz to 1.05 THz. Special types of network analyzers can also cover lower frequency ranges down to 1 Hz. These network analyzers can be used for example for the stability analysis of open loops or for the measurement of audio and ultrasonic components.

The two basic types of network analyzers are:

- Scalar network analyzer (SNA)—measures amplitude properties only.
- Vector network analyzer (VNA)—measures both amplitude and phase properties.

A VNA is a form of RF network analyzer widely used for RF design applications. A VNA may also be called a gain-phase meter or an automatic network analyzer. An SNA is functionally identical to a spectrum analyzer in combination with a tracking generator. As of 2007[update], VNAs are the most common type of network analyzers, and so references to an unqualified "network analyzer" most often mean a VNA. Three prominent VNA manufacturers are Keysight, Anritsu, and Rohde & Schwarz.

Another category of network analyzer is the microwave transition analyzer (MTA) or large signal network analyzer (LSNA), which measure both amplitude and phase of the fundamental and harmonics. The MTA was commercialized before the LSNA, but was lacking some of the user-friendly calibration features now available with the LSNA.

Architecture

The basic architecture of a network analyzer involves a signal generator, a test set, one or more receivers and display. In some setups, these units are distinct instruments. Most VNAs have two test ports, permitting measurement of four S-parameters (S_{11}, S_{21}, S_{12} and S_{22}), but instruments with more than two ports are available commercially.

Signal Generator

The network analyzer needs a test signal, and a signal generator or signal source will provide one. Older network analyzers did not have their own signal generator, but had the ability to control a stand-alone signal generator using, for example, a GPIB connection. Nearly all modern network analyzers have a built-in signal generator. High-performance network analyzers have two built-in sources. Two built-in sources are useful for applications such as mixer test, where one source provides the RF signal, another the LO, or amplifier intermodulation testing, where two tones are required for the test.

Test Set

The test set takes the signal generator output and routes it to the device under test, and it routes the signal to be measured to the receivers. It often splits off a reference channel for the incident wave. In a SNA, the reference channel may go to a diode detector (receiver) whose output is sent to the signal generator's automatic level control. The result is better control of the signal generator's output and better measurement accuracy. In a VNA, the reference channel goes to the receivers; it is needed to serve as a phase reference.

Directional couplers or two resistor power dividers are used for signal separation. Some microwave test sets include the front end mixers for the receivers (e.g., test sets for HP 8510).

Receiver

The receivers make the measurements. A network analyzer will have one or more receivers connected to its test ports. The reference test port is usually labeled R, and the primary test ports are A, B, C. Some analyzers will dedicate a separate receiver to each test port, but others share one or two receivers among the ports. The R receiver may be less sensitive than the receivers used on the test ports.

For the SNA, the receiver only measures the magnitude of the signal. A receiver can be a detector diode that operates at the test frequency. The simplest SNA will have a single test port, but more accurate measurements are made when a reference port is also used. The reference port will compensate for amplitude variations in the test signal at the measurement plane. It is possible to share a single detector and use it for both the reference port and the test port by making two measurement passes.

For the VNA, the receiver measures both the magnitude and the phase of the signal. It needs a reference channel (R) to determine the phase, so a VNA needs at least two receivers. The usual method down converts the reference and test channels to make the measurements at a lower frequency. The phase may be measured with a quadrature detector. A VNA requires at least two receivers, but some will have three or four receivers to permit simultaneous measurement of different parameters.

There are some VNA architectures (six-port) that infer phase and magnitude from just power measurements.

Processor and Display

With the processed RF signal available from the receiver / detector section it is necessary to display the signal in a format that can be interpreted. With the levels of processing that are available today, some very sophisticated solutions are available in RF network analyzers. Here the reflection and transmission data is formatted to enable the information to be interpreted as easily as possible. Most RF network analyzers incorporate features including linear and logarithmic sweeps, linear and log formats, polar plots, Smith charts, etc. Trace markers, limit lines and also pass / fail criteria may also be added in many instances.

S-parameter Measurement with Vector Network Analyzer

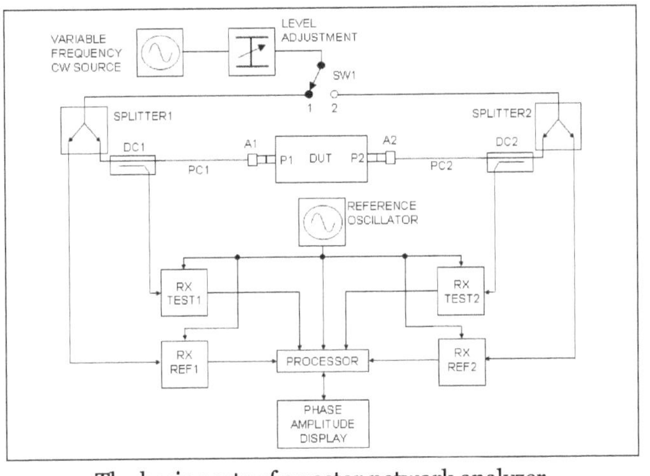

The basic parts of a vector network analyzer.

A VNA is a test system that enables the RF performance of radio frequency and microwave devices to be characterised in terms of network scattering parameters, or S parameters.

The diagram shows the essential parts of a typical 2-port vector network analyzer (VNA). The two ports of the device under test (DUT) are denoted port 1 (P1) and port 2 (P2). The test port connectors provided on the VNA itself are precision types which will normally have to be extended and connected to P1 and P2 using precision cables 1 and 2, PC1 and PC2 respectively and suitable connector adaptors A1 and A2 respectively.

The test frequency is generated by a variable frequency CW source and its power level is set using a variable attenuator. The position of switch SW1 sets the direction that the test signal passes through the DUT. Initially consider that SW1 is at position 1 so that the test signal is incident on the DUT at P1 which is appropriate for measuring S_{11} and S_{21}. The test signal is fed by SW1 to the common port of splitter 1, one arm (the reference channel) feeding a reference receiver for P1 (RX REF1) and the other (the test channel) connecting to P1 via the directional coupler DC1, PC1 and A1. The third port of DC1 couples off the power reflected from P1 via A1 and PC1, then feeding it to test receiver 1 (RX TEST1). Similarly, signals leaving P2 pass via A2, PC2 and DC2 to RX TEST2. RX REF1, RX TEST1, RX REF2 and RXTEST2 are known as coherent receivers as they share the same reference oscillator, and they are capable of measuring the test signal's amplitude *and* phase at the test frequency. All of the complex receiver output signals are fed to a processor which does the mathematical processing and displays the chosen parameters and format on the phase and amplitude display. The instantaneous value of phase includes both the temporal and spatial parts, but the former is removed by virtue of using 2 test channels, one as a reference and the other for measurement. When SW1 is set to position 2, the test signals are applied to P2, the reference is measured by RX REF2, reflections from P2 are coupled off by DC2 and measured by RX TEST2 and signals leaving P1 are coupled off by DC1 and measured by RX TEST1. This position is appropriate for measuring S_{22} and S_{12}.

Calibration and Error Correction

A network analyzer, like most electronic instruments requires periodic calibration - typically this is performed once per year and is performed by the manufacturer or by a 3rd party in a calibration laboratory. When the instrument is calibrated, it will usually have a sticker fixed to the outside, stating the date it was calibrated and when the next calibration is due. A calibration certificate will be issued.

A vector network analyzer achieves highly accurate measurements by correcting for the systematic errors in the instrument, the characteristics of cables, adapters and test fixtures. The process of error correction, although commonly just called calibration, is an entirely different process, and may be performed by an engineer several times in an hour. Sometimes it is called user-calibration, to indicate the difference from periodic calibration by a manufacturer.

A network analyzer has connectors on its front panel, but the measurements are seldom made at the front panel. Usually some test cables will connect from the front panel to the device under test (DUT). The length of those cables will introduce a time delay and corresponding phase shift (affecting VNA measurements); the cables will also introduce some attenuation (affecting SNA and VNA measurements). The same is true for cables and couplers inside the network analyzer. All these factors will change with temperature. Calibration usually involves measuring known standards and using those measurements to compensate for systematic errors, but there are methods which do not require known standards. Only systematic errors can be corrected. Random errors, such as connector repeatability can not be corrected by the user calibration. However, some portable vector network analyzers, designed for lower accuracy measurement outside using batteries, do attempt some correction for temperature by measuring the internal temperature of the network analyzer.

The first steps, prior to actually starting the user calibration are:

- Visually inspect the connectors for any problems such as bent pins or parts which are obviously off-centre. These should be thrown away, as mating damaged connectors with good connectors will often result in damaging the good connector.
- Clean the connectors with compressed air at less than 60 psi.
- If necessary clean the connectors with isopropyl alcohol.
- Gage the connectors to determine there are not any gross mechanical problems. Connector gauges with resolutions of 0.001" to 0.0001" will usually be included in the better quality calibration kits.
- Torque the connectors to the specified torque. A torque wrench will be supplied with all but the cheapest calibration kits.

There are several different methods of calibration:

- SOLT: Which is an acronym for Short, Open, Load, Thru, is the simplest method. As the name suggests, this requires access to known standards with a short circuit, open circuit, a precision load (usually 50 ohms) and a through connection. It is best if the test ports have the same type of connector (N, 3,5 mm etc.), but of a different sex, so the through just requires the test ports are connected together. SOLT is suitable for coaxial measurements, where it is possible to obtain the short, open, load and through. The SOLT calibration method is less suitable for waveguide measurements, where it is difficult to obtain an open circuit or a load, or for measurements on non-coaxial test fixtures, where the same problems with finding suitable standards exist.

- TRL (through-reflect-line calibration): This technique is useful for microwave, noncoaxial environments such as fixture, wafer probing, or waveguide. TRL uses a transmission line, significantly longer in electrical length than the through line, of known length and impedance as one standard. TRL also requires a high-reflection standard (usually, a short or open) whose impedance does not have to be well characterized, but it must be electrically the same for both test ports.

The simplest calibration that can be performed on a network analyzer is a transmission measurement. This gives no phase information, and so gives similar data to a scalar network analyzer. The simplest calibration that can be performed on a network analyzer, whilst providing phase information is a 1-port calibration (S11 or S22, but not both). This accounts for the three systematic errors which appear in 1-port reflectivity measurements:

- Directivity—error resulting from the portion of the source signal that never reaches the DUT.
- Source match—errors resulting from multiple internal reflections between the source and the DUT.
- Reflection tracking—error resulting from all frequency dependence of test leads, connections, etc.

In a typical 1-port reflection calibration, the user measures three known standards, usually an open, a short and a known load. From these three measurements the network analyzer can account for the three errors above.

A more complex calibration is a full 2-port reflectivity and transmission calibration. For two ports there are 12 possible systematic errors analogous to the three above. The most common method for correcting for these involves measuring a short, load and open standard on each of the two ports, as well as transmission between the two ports.

It is impossible to make a perfect short circuit, as there will always be some inductance in the short. It is impossible to make a perfect open circuit, as there will always be some fringing capacitance. A modern network analyzer will have data stored about the devices in a calibration kit. (Keysight Technologies 2006) For the open-circuit, this will be some electrical delay (typically tens of picoseconds), and fringing capacitance which will be frequency dependent. The capacitance is normally specified in terms of a polynomial, with the coefficients specific to each standard. A short will have some delay, and a frequency dependent inductance, although the inductance is normally considered insignificant below about 6 GHz. The definitions of the standards for a particular calibration kit will often change depending on the frequency range of the network analyzer. If a calibration kit works to 9 GHz, but a particular network analyzer has a maximum frequency of operation of 3 GHz, then the capacitance of the open standard can approximated more closely up to 3 GHz, using a different set of coefficients than are necessary to work up to 9 GHz.

In some calibration kits, the data on the males is different from the females, so the user needs to specify the sex of the connector. In other calibration kits (e.g. Keysight 85033E 9 GHz 3.5 mm), the male and female have identical characteristics, so there is no need for the user to specify the sex. For sexless connectors, like APC-7, this issue does not arise.

Most network analyzers will have the ability to have a user defined calibration kit. So if a user has a particular calibration kit, details of which are not in the firmware of the network analyzer, the data about the kit can be loaded into the network analyzer and so the kit used. Typically the calibration data can be entered on the instrument front panel, as well as loaded from a medium such as floppy disk or USB stick, or down a bus such as USB or GPIB.

The more expensive calibration kits will usually include a torque wrench to tighten connectors properly and a connector gauge to ensure there are no gross errors in the connectors.

Automated Calibration Fixtures

A calibration using a mechanical calibration kit may take a significant amount of time. Not only must the operator sweep through all the frequencies of interest, but the operator must also disconnect and reconnect the various standards. To avoid that work, network analyzers can employ automated calibration standards. The operator connects one box to the network analyzer. The box has a set of standards inside and some switches that have already been characterized. The network analyzer can read the characterization and control the configuration using a digital bus such as USB.

Network Analyzer Verification Kits

A number of verification kits are available to verify the network analyzer is performing to specification. These typically consist of transmission lines with an air dielectric and attenuators. The Agilent 85055A kit includes a 10 cm airline, stepped impedance airline, 20 dB and 50 dB attenuators with data on the devices measured by the manufacturer and stored on both a floppy disk and USB flash drive. Older versions of the 85055A have the data stored on tape and floppy disks rather than on USB drives.

Noise Figure Measurements

The three major manufacturers of VNAs, Keysight, Anritsu, and Rohde & Schwarz, all produce models which permit the use of noise figure measurements. The vector error correction permits higher accuracy than is possible with other forms of commercial noise figure meters.

Megger

The Megger is the instrument which is used for measuring the resistance of the insulation. It works on the principle of comparison, i.e., the resistance of the insulation is compared with the known value of resistance. If the resistance of the insulation is high, the pointer of the moving coil deflects towards the infinity, and if it is low, then the pointer indicates zero resistance. The accuracy of the Megger is high as compared to other instruments.

Insulation resistance IR quality of an electrical system degrades with time, environment condition, i.e., temperature, humidity, moisture and dust particles. It also gets impacted negatively due to the presence of electrical and mechanical stress, so it's become very necessary to check the IR (Insulation resistance) of equipment at a constant regular interval to avoid any measure fatal or electrical shock.

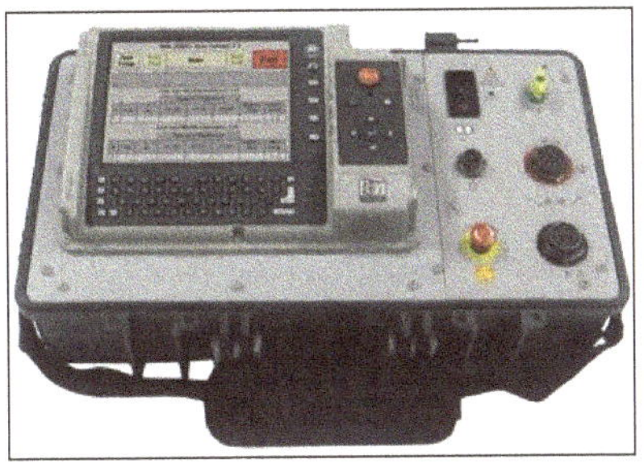

Uses of Megger

The device enables us to measure electrical leakage in wire, results are very reliable as we shall be passing electric current through device while we are testing. The equipment basically uses for verifying the electrical insulation level of any device such as motors, cables, generators, windings, etc. This is a very popular test being carried out since very long back. Not necessary it shows us exact area of electrical puncture but shows the amount of leakage current and level of moisture within electrical equipment/winding/system.

Types of Megger

This can be separated into mainly two categories:

- Electronic Type (Battery Operated).
- Manual Type (Hand Operated).

But there are another types of megger which is motor operated type which does not use battery to produce voltage it requires external source to rotate a electrical motor which in turn rotates the generator of the megger.

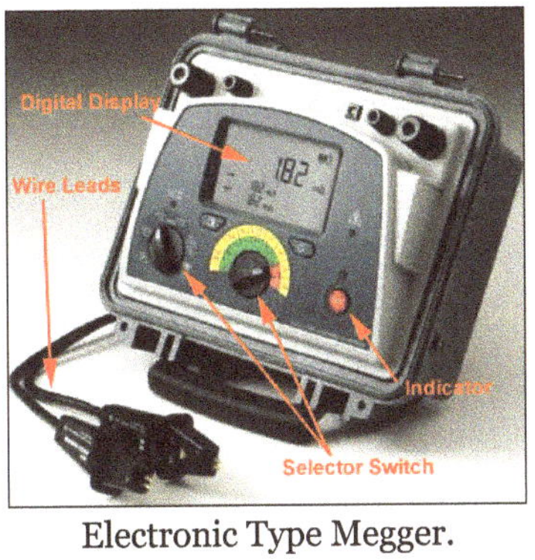

Electronic Type Megger.

Electronic Type Megger

Important parts:

- Digital Display: A digital display to show IR value in digital form.
- Wire Leads: Two nos of wire leads for connecting megger with electrical external system to be tested.
- Selection Switches: Switches use to select electrical parameters ranges.
- Indicators: To indicates various parameters status i.e. On-Off. For Example Power, hold, Warning, etc.

Advantages of Electronic Type Megger

- Level of accuracy is very high.
- IR value is digital type, easy to read.
- One person can operate very easily.
- Works perfectly even at very congested space.
- Very handy and safe to use.

Disadvantages of Electronic Type Megger

- Require an external source of energy to energies i.e. Dry cell.
- Costlier in market.

Hand Operated Megger

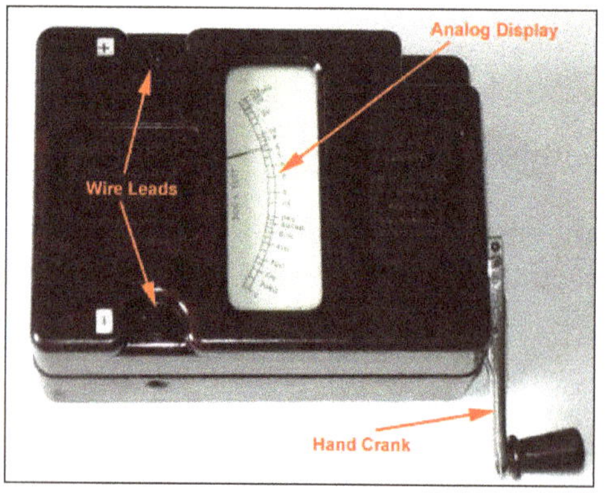

Important parts:

- Analog display: Analog display provided on front face of tester for IR value recording.
- Hand Crank: Hand crank used to rotate helps to achieve desired RPM required generate voltage which runs through electrical system.
- Wire Leads: Used same as in electronic tester i.e. For connecting tester with electrical system.

Advantages of Hand Operated Megger

- Still keeps important in such high-tech world as it's an oldest method for IR value determination.
- No external source required to operate.
- Cheaper available in market.

Disadvantages of Hand Operated Megger

- At least 2 person required to operate i.e. one for rotation of crank other to connect megger with electrical system to be tested.
- Accuracy is not up to the level as it's varies with rotation of crank.
- Require very stable placement for operation which is a little hard to find at working sites.
- Unstable placement of tester may impact the result of tester.
- Provides an analog display result.
- Require very high care and safety during use of the same.

Construction of Megger

Circuit Construction features:

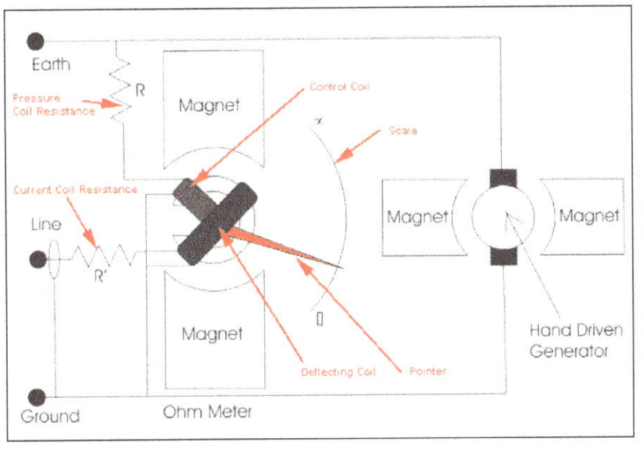

- Deflecting and Control coil: Connected parallel to the generator, mounted at right angle to each other and maintain polarities in such a way to produced torque in opposite direction.
- Permanent Magnets: Produce magnetic field to deflect pointer with North-South pole magnet.
- Pointer: One end of the pointer connected with coil another end deflects on scale from infinity to zero.
- Scale: A scale is provided in front-top of the megger from range 'zero' to 'infinity', enable us to read the value.
- D.C generator or Battery connection: Testing voltage is produced by hand operated DC

generator for manual operated Megger. Battery / electronic voltage charger is provided for automatic type Megger for same purpose.

- Pressure Coil Resistance and Current Coil Resistance: Protect instrument from any damage because of low external electrical resistance under test.

Working Principle of Megger

- Voltage for testing produced by hand operated megger by rotation of crank in case of hand operated type, a battery is used for electronic tester.
- 500 Volt DC is sufficient for performing test on equipment range up to 440 Volts.
- 1000 V to 5000 V is used for testing for high voltage electrical systems.
- Deflecting coil or current coil connected in series and allows flowing the electric current taken by the circuit being tested.
- The control coil also known as pressure coil is connected across the circuit.
- Current limiting resistor (CCR and PCR) connected in series with control and deflecting coil to protect damage in case of very low resistance in external circuit.
- In hand operated megger electromagnetic induction effect is used to produce the test voltage i.e. armature arranges to move in permanent magnetic field or vice versa.
- Where as in electronic type megger battery are used to produce the testing voltage.
- As the voltage increases in external circuit the deflection of pointer increases and deflection of pointer decreases with a increases of current.
- Hence, resultant torque is directly proportional to voltage and inversely proportional to current.
- When electrical circuit being tested is open, torque due to voltage coil will be maximum and pointer shows 'infinity' means no shorting throughout the circuit and has maximum resistance within the circuit under test.
- If there is short circuit pointer shows 'zero', which means 'NO' resistance within circuit being tested.

Work philosophy based on ohm-meter or ratio-meter. The deflection torque is produced with megger tester due to the magnetic field produced by voltage and current, similarly like 'Ohm's Law'. The torque of the megger varies in a ration with V/I, (Ohm's Law:- $V = IR$ or $R = V/I$). Electrical resistance to be measured is connected across the generator and in series with deflecting coil. Produced torque shall be in opposite direction if current supplied to the coil.

- High Resistance = No Current: No current shall flow through deflecting coil, if resistance is very high i.e. infinity position of pointer.
- Small Resistance = High Current: If circuit measures small resistance allows a high electric

current to pass through deflecting coil, i.e. produced torque make the pointer to set at 'ZERO'.

- Intermediate Resistance = Varied Current: If measured resistance is intermediate, produced torque align or set the pointer between the range of 'ZERO to INIFINITY'.

Connection Diagram of Megger for Testing

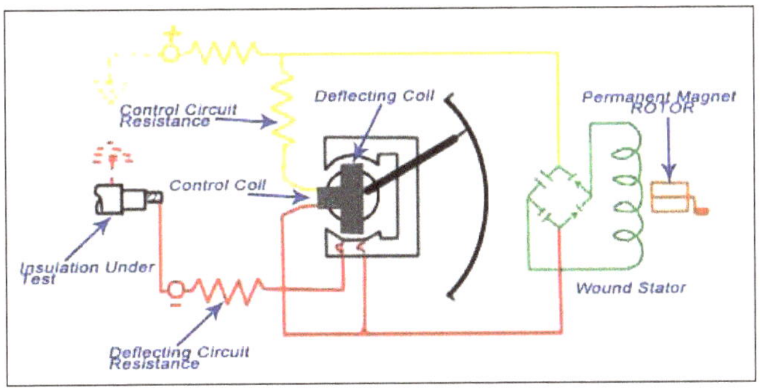

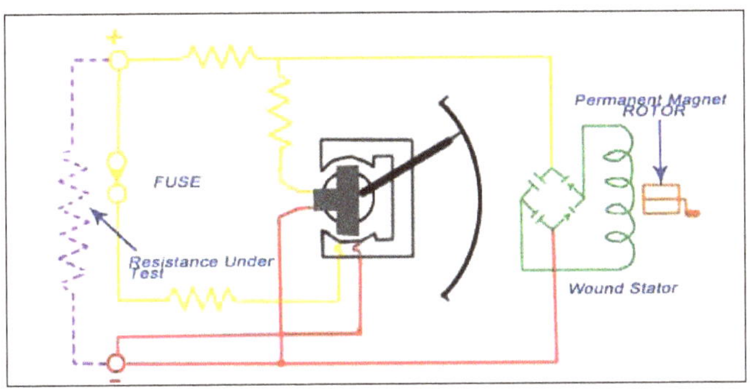

Power Factor Meter

The power factor meter measures the power factor of a transmission system. The power factor is the cosine of the angle between the voltage and current. The power factor meter determines the types of load using on the line, and it also calculates the losses occur on it.

The power factor of the transmission line is measured by dividing the product of voltage and current with the power. And the value of voltage current and power is easily determined by the voltmeter, ammeter and wattmeter respectively. This method gives high accuracy, but it takes time.

The power factor of the transmission line is continuously changed with time. Hence it is essential to take the quick reading. The power factor meter takes a direct reading, but it is less accurate. The reading obtained from the power factor meter is sufficient for many purposes to expect precision testing.

The power factor meter has the moving system called pointer which is in equilibrium with the two opposing forces. Thus, the pointer of the power factor meter remains at the same position which is occupied by it at the time of disconnection.

Types of Power Factor Meter

The power factor meter is of two types. They are:

- Electrodynamometer:
    - Single Phase Electrodynammeter.
    - Three Phases Electrodynamometer.
- Moving Iron Type Meter:
    - Rotating Iron Magnetic Field.
    - Number of Alternating Field.

The different types of power factor meter are explained below in details.

Single Phase Electrodynamometer Power Factor Meter

The construction of the single phase electrodynamometer is shown in the figure below. The meter has fixed coil which acts as a current coil. This coil is split into two parts and carry the current under test. The magnetic field of the coil is directly proportional to the current flow through the coil.

The meter has two identical pressure coils A and B. Both the coils are pivoted on the spindle. The pressure coil A has no inductive resistance connected in series with the circuit, and the coil B has highly inductive coil connected in series with the circuit.

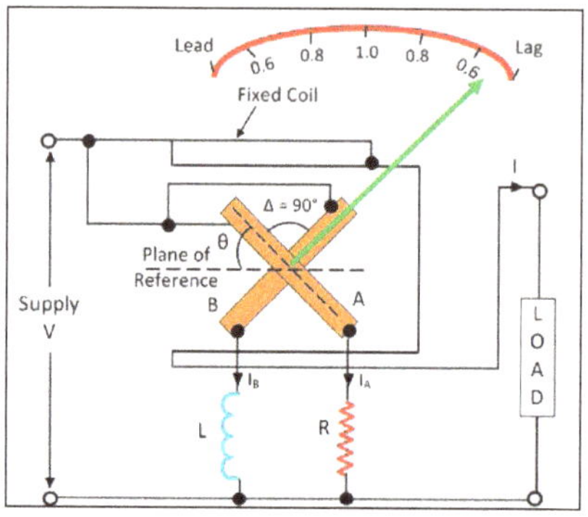

Single phase electrodynamometer type power factor meter.

The current in the coil A is in phase with the circuit while the current in the coil B lag by the voltage nearly equal to 90°. The connection of the moving coil is made through silver or gold ligaments which minimize the controlling torque of the moving system.

The meter has two deflecting torque one acting on the coil A, and the other is on coil B. The windings are so arranged that they are opposite in directions. The pointer is in equilibrium when the torques are equal.

Deflecting torque acting on the coil A is given as,

$$T_A = KVIM \cos\phi \sin\theta$$

θ – angular deflection from the plane of reference.

M_{max} – maximum value of mutual inductance between the coils.

The deflecting torque acting on coil B is expressed as:

$$I_B = KVIM_{max} Cos(90^{\circ} - \phi) Sin(90^{\circ} + \phi)$$
$$I_B = KVIM_{max} \cos\phi \sin\theta$$

The deflecting torque is acting on the clockwise direction. The value of maximum mutual inductance is same between both the deflecting equations.

$$T_A = T_B$$
$$KIVM \cos\phi \sin\theta = KIVM_{max} \cos\phi \sin\theta$$

This torque acts on anti-clockwise direction. The above equation shows that the deflecting torque is equal to the phase angle of the circuit.

Three Phase Electrodynometer Power Factor Meter

The construction of the three phase meter is shown in the figure below. The electrodynamometer is only useful for the balanced load. The moving coil is placed at an angle of 120°. They are connected across different phases of the supply circuit. Both the coil has a series resistance.

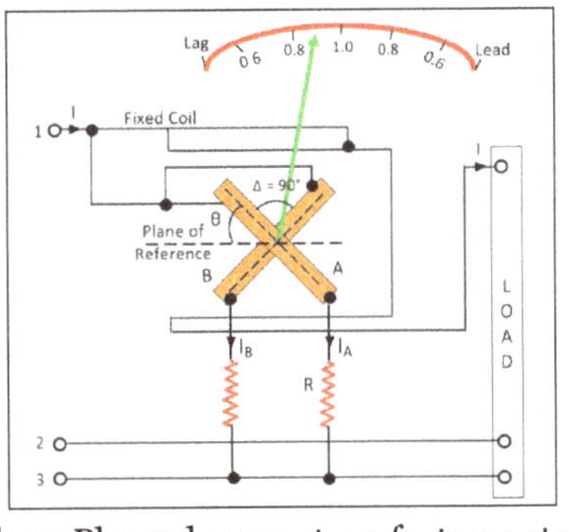

Three Phase dynamo type factor meter.

The voltage across the coil A is V_{12} and the current across it I_{A1}. The circuit of the coil is resistive, and hence the current and voltage are in phase with each other. Similarly, the voltage V_{13} and the current I_{B1} is in phase with each other. The phasor diagram of the three phase electrodynamic meter is shown in the figure below.

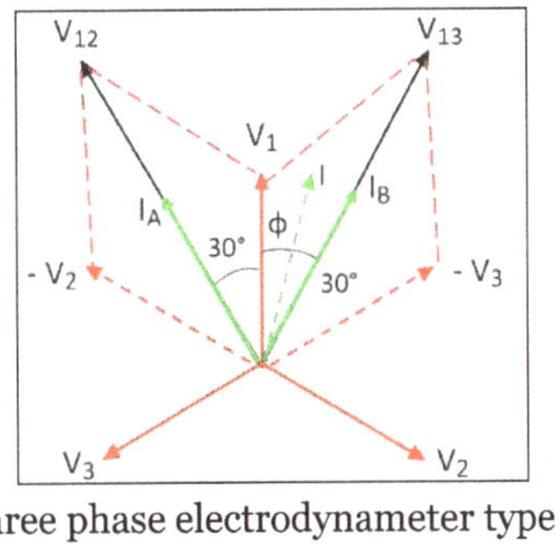

Phasor diagram of three phase electrodynameter type power factor meter.

Let Φ – phase angle of the circuit.

θ – angular deflection from the plane of reference.

Torque acting on coil A is

$$T_A = KIV_{12}M_{\max}Cos(30^\circ + \phi)\sin(60^\circ + \phi)$$
$$T_A = \sqrt{3}KIV_{12}M_{\max}Cos(30^\circ + \phi)\sin(60^\circ + \phi)$$

Torque acting on coil B is,

$$T_B = KIV12M_{\max}Cos(30^\circ - \phi)Sin(120^\circ + \phi)$$
$$T_B = KIV12M_{\max}Cos(30^\circ - \phi)Sin(120^\circ + \phi)$$

The torque T_A and T_B are acting on the opposite directions.

$$Cos(30^\circ - \phi)Sin(120^\circ + \phi) = Cos(30^\circ - \phi)Sin(120^\circ + \phi)$$

Thus the angular deflection of the coil is directly proportional to the phase angle of the circuit.

Moving Iron Power Factor Meter

The moving iron instrument is divided into two categories. They are the rotating magnetic field to some alternating fields.

A. Rotating Field Power factor Meter – The following are the essential feature of the rotating magnetic field. The power factor meter has three fixed coils, and their axes are 120⁰ displaced from each other. The axes are intersecting each other. The coils are connected to the three phase supply with the help of the current transformer.

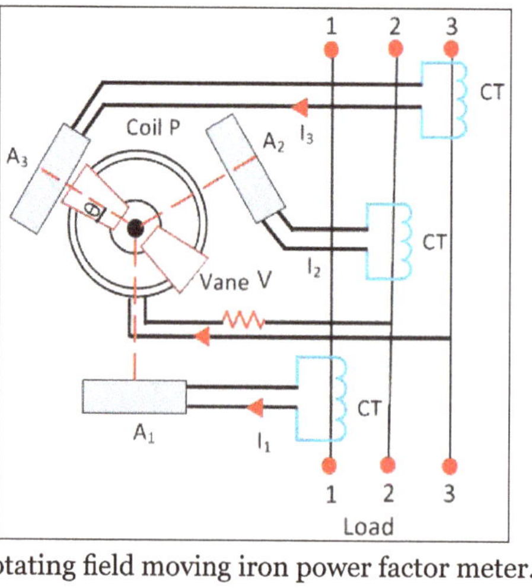

Rotating field moving iron power factor meter.

The P is the fixed coil connected in series with the high resistance circuit across the phases 2 and 3. There is an iron cylinder across coil P. The two iron vanes are fixed to the cylinder. The spindles also carry damping vanes and pointer.

The phasor diagram of the power factor meter is shown in the figure.

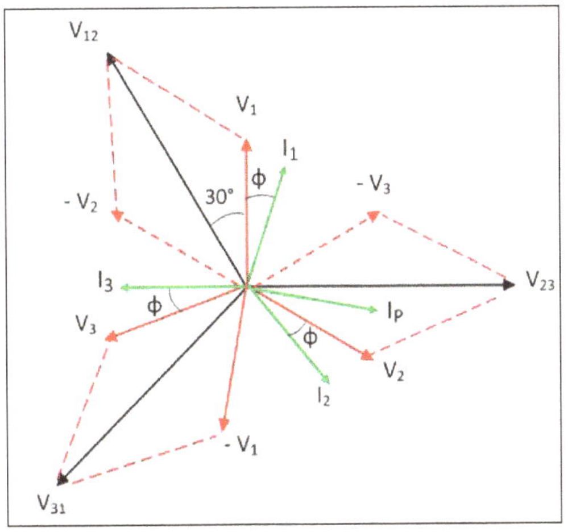

Phasor diagram for moving iron type power factor meter.

The total torque of the meter is zero for steady state deflection.

$$[Cos(90^\circ - \phi)Sin(90^\circ + \phi) + Cos(330^\circ - \phi)Sin(210^\circ + \phi) + Cos(210^\circ + \phi)Sin(330^\circ + \phi) = 0$$

The coil P and the iron cylinders generate the alternating flux which interacts with the flux of the fixed coils. The interaction of the coil generates the moving system which determined the phase angle of the current. The vanes of the power factor meter are magnetized by the current of the moving coil which is in phase with the system line voltage.

Advantages of Moving Iron Power Factor

- The meter requires large working force as compared to the electrodynamometer type meter.
- The coils of the moving iron instruments are fixed permanently.
- The range of the scale extends up to 360°.
- The construction of the meter is robust and simple.
- The moving iron instrument is cheap as compared to electrodynamic meter.

Disadvantages of Moving Iron Instrument

- The loss occurs in the iron part of the meter. The losses depend on the load and the frequency of the meter.
- The meter has low accuracy.
- The calibration of the meter is affected because of the variation in supply frequencies, voltage and waveforms etc.

The power factor meter is used for measuring the power factor of the balanced load.

Wattmeter

The wattmeter is an instrument for measuring the electric power (or the supply rate of electrical energy) in watts of any given circuit. Electromagnetic wattmeters are used for measurement of utility frequency and audio frequency power; other types are required for radio frequency measurements.

Electrodynamic

Early wattmeter on display at the Historic Archive and Museum of Mining.

The traditional analog wattmeter is an electrodynamic instrument. The device consists of a pair of fixed coils, known as *current coils*, and a movable coil known as the *potential coil.*

The current coils are connected in series with the circuit, while the potential coil is connected in parallel. Also, on analog wattmeters, the potential coil carries a needle that moves over a scale to indicate the measurement. A current flowing through the current coil generates an electromagnetic field around the coil. The strength of this field is proportional to the line current and in phase with it. The potential coil has, as a general rule, a high-value resistor connected in series with it to reduce the current that flows through it.

The result of this arrangement is that on a DC circuit, the deflection of the needle is proportional to *both* the current (I) *and* the voltage (V), thus conforming to the equation $P=VI$.

For AC power, current and voltage may not be in phase, owing to the delaying effects of circuit inductance or capacitance. On an AC circuit the deflection is proportional to the average instantaneous product of voltage and current, thus measuring active power, $P=VI \cos \varphi$. Here, $\cos\varphi$ represents the power factor which shows that the power transmitted may be less than the apparent power obtained by multiplying the readings of a voltmeter and ammeter in the same circuit.

The two circuits of a wattmeter can be damaged by excessive current. The ammeter and voltmeter are both vulnerable to overheating — in case of an overload, their pointers will be driven off scale — but in the wattmeter, either or even both the current and potential circuits can overheat *without* the pointer approaching the end of the scale. This is because the position of the pointer depends on the power factor, voltage and current. Thus, a circuit with a low power factor will give a low reading on the wattmeter, even when both of its circuits are loaded to the maximum safety limit. Therefore, a wattmeter is rated not only in watts, but also in volts and amperes.

A typical wattmeter in educational labs has two voltage coils (pressure coils) and a current coil.

The two pressure coils can be connected in series or parallel to change the ranges of the wattmeter. The pressure coil can also be tapped to change the meter's range. If the pressure coil has range of 300 volts, the half of it can be used so that the range becomes 150 volts.

Electronic

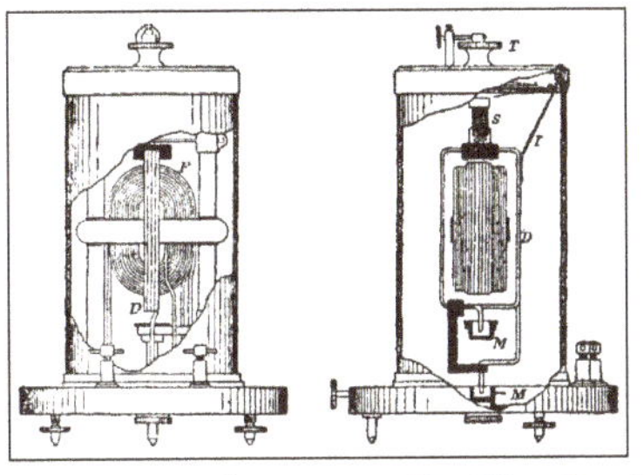

Siemens electrodynamometer, circa 1910, F = Fixed coil, D = Movable coil, S = Spiral spring, T = Torsion head, M = Mercury cups, I = Index needle.

Prodigit Model 2000MU (UK version), shown in use and displaying a reading of 10 watts being consumed by the appliance.

Electronic wattmeters are used for direct, small power measurements or for power measurements at frequencies beyond the range of electrodynamometer-type instruments.

Digital

A modern digital wattmeter samples the voltage and current thousands of times a second. For each sample, the voltage is multiplied by the current at the same instant; the average over at least one cycle is the real power. The real power divided by the apparent volt-amperes (VA) is the power factor. A computer circuit uses the sampled values to calculate RMS voltage, RMS current, VA, power (watts), power factor, and kilowatt-hours. The readings may be displayed on the device, retained to provide a log and calculate averages, or transmitted to other equipment for further use. Wattmeters vary considerably in correctly calculating energy consumption, especially when real power is much lower than VA (highly reactive loads, e.g. electric motors). Simple meters may be calibrated to meet specified accuracy only for sinusoidal waveforms. Waveforms for switched-mode power supplies as used for much electronic equipment may be very far from sinusoidal, leading to unknown and possibly large errors at any power. This may not be specified in the meter's manual.

Precision and Accuracy

There are limitations to measuring power with inexpensive wattmeters, or indeed with any meters not designed for low-power measurements. This particularly affects low power (e.g. under

10 watts), as used in standby; readings may be so inaccurate as to be useless (although they do confirm that standby power is low, rather than high). The difficulty is largely due to difficulty in accurate measurement of the alternating current, rather than voltage, and the relatively little need for low-power measurements. The specification for the meter should specify the reading error for different situations. For a typical plug-in meter the error in wattage is stated as ±5% of measured value ±10 W (e.g., a measured value of 100W may be wrong by 5% of 100 W plus 10 W, i.e., ±15 W, or 85–115 W); and the error in kW·h is stated as ±5% of measured value ±0.1 kW·h. If a laptop computer in sleep mode consumes 5 W, the meter may read anything from 0 to 15.25 W, without taking into account errors due to non-sinusoidal waveform. In practice accuracy can be improved by connecting a fixed load such as an incandescent light bulb, adding the device in standby, and using the difference in power consumption. This moves the measurement out of the problematic low-power zone.

Radio Frequency

Instruments with moving coils can be calibrated for direct current or power frequency currents up to a few hundred hertz. At radio frequencies a common method is a rectifier circuit arranged to respond to current in a transmission line; the system is calibrated for the known circuit impedance. Diode detectors are either directly connected to the source, or used with a sampling system that diverts only a portion of the RF power through the detector. Thermistors and thermocouples are used to measure heat produced by RF power and can be calibrated either directly or by comparison with a known reference source of power. A bolometer power sensor converts incident radio frequency power to heat. The sensor element is maintained at a constant temperature by a small direct current. The reduction in current required to maintain temperature is related to the incident RF power. Instruments of this type are used throughout the RF spectrum and can even measure visible light power. For high-power measurements, a calorimeter directly measures heat produced by RF power.

Watthour Meters

Itron OpenWay wattmeter with two-way communications for remote reading, in use by DTE Energy.

An instrument which measures electrical energy in watt hours is essentially a wattmeter which accumulates or averages readings. Digital electronic instruments measure many parameters and can be used where a wattmeter is needed: volts, current,in amperes, apparent instantaneous power, actual power, power factor, energy in [k]W·h over a period of time, and cost of electricity consumed.

Electrodynamometer Wattmeter

The instrument whose working depends on the reaction between the magnetic field of moving and fixed coils is known as the Electrodynamo-meter Wattmeter. It uses for measuring the power of both the AC and DC circuits.

The working principle of the Electrodynamometer Wattmeter is very simple and easy. Their working depends on the theory that the current carrying conductor placed in a magnetic field experiences a mechanical force. This mechanical force deflects the pointer which is mounted on the calibrated scale.

Construction of Electrodynamometer Wattmeter

The following are the important parts of the Electrodynamometer Wattmeter:

- Fixed coil: The fixed coil connects in series with the load. It is considered as a current coil because the load current flows through it. For making the construction easy the fixed coil divide into two parts. And these two elements are parallel connected to each other. The fixed coil produces the uniform electric field which is essentials for the working of the instruments. The current coil of the instruments is designed to carry the current of approximately 20 amperes for saving the power.
- Moving Coil: The moving coil consider as the pressure coil of the instruments. It connects in parallel with the supply voltage. The current flows through them is directly proportional to the supply voltage. The pointer mounts on the moving coil. The movement of the pointer controls with the help of the spring. The current flows through the coil increases their temperature. The flows of currents control with the help of resistor which connects in series with the moving coil.
- Control: The control system provides the controlling torque to the instruments. The gravity control and the spring control are the two types of control system. Out of two, the Electrodynamometer Wattmeter uses spring control system. The spring control system is used for the movement of the pointer.
- Damping: The damping is the effect which reduces the movement of the pointer. In this Wattmeter the damping torque produces because of the air friction. The other types of damping are not used in the system because they destroy the useful magnetic flux.
- Scales and pointers: The instruments use a linear scale because their moving coil moves linearly. The apparatus uses the knife edge pointer for removing the parallax error which causes because of oversights.

Working of Electrodynamometer Wattmeter

The Electrodynamometer Wattmeter has two types of coils; fixed and the moving coil. The fixed coil connects in series with the circuit whose power consumption use to be measured. The supply voltage applies to the moving coil. The resistor controls the current across the moving coil, and it is connected in series with it.

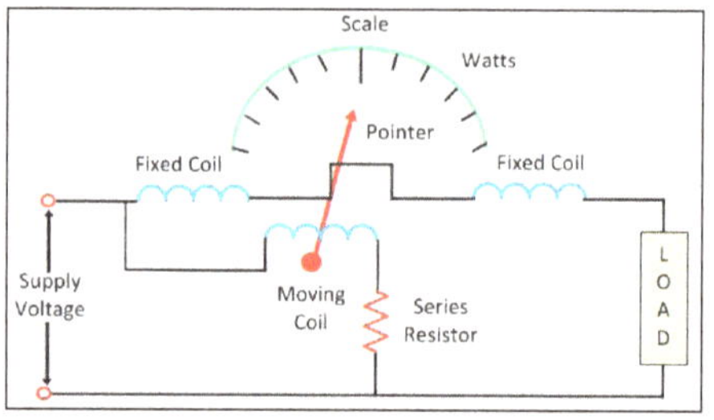

Electrodynameter Wattmeter.

The pointer is fixed on the moving coil which is placed between the fixed coils. The current and voltage of the fixed and moving coil generate the two magnetic fields. And the interaction of these two magnetic fields deflects the pointer of the instrument. The deflection of the pointer is directly proportional to the power flows through it.

Theory of Electrodynamometer Wattmeter

The circuit diagram of the electrodynamometer wattmeter is shown in the figure below.

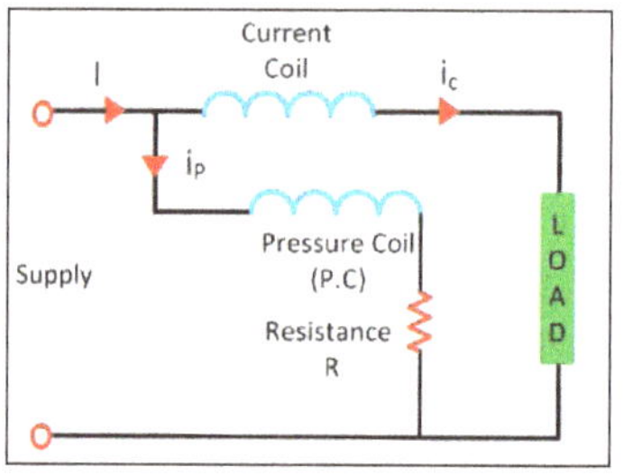

Circuit of electrodynamo wattmeter.

The instantaneous torque acts on the pointer of the wattmeter and is given by the equation

$$T_1 = i_1 i_p \frac{dM}{d\theta}$$

Where,

- i_p – pressure coil current.
- i_c– current coil current.
- dm/dθ – the rate of change of deflection of pointer concerning angle θ.

The voltage across the pressure coil of the circuit is given as

$$v = \sqrt{2I} \sin(\omega t - \phi)$$

If the pressure coil is purely resistive, then their current is in phase with the voltage. And the value of current is given by the equation.

$$i_p = \frac{v}{R_p} = \sqrt{2}\left(\frac{VI}{R_p}\right) \text{Sin } \omega t = \sqrt{2} I_p \text{ Sin } \omega t$$

If the current coil lag by a voltage in phase angle Φ, the current through the current coil is given as:

$$i_p = \sqrt{2} I \sin(\omega t - \phi)$$

The value of the current in the pressure coil is very small. Hence the current flows through the pressure coil is considered as the total load current. The torque acts on the coils becomes,

$$T_i = \sqrt{2} I_p \operatorname{Sin}(\omega t - \phi) \frac{dM}{d\theta}$$

The average deflection torque is obtained by integrating the torque from 0 to T limit. The average deflection torque of the coil is given as,

$$Ti = \sqrt{2}(\frac{VI}{R_p}) \cos\phi . \frac{dM}{d\theta}$$

The controlling torque exerted on the spring is given by,

$$T_c = K\theta$$

Errors in Electrodynamometer Wattmeter

The following are the errors in the Electrodynamometer Wattmeter:

- Pressure Coil Inductance: The pressure coil of the Electrodynamometer has some inductance. Because of the inductance, the current of the pressure coils lags behind the voltage. Thus, the power factor of the wattmeter becomes lagging, and the meter reads high reading.
- Pressure Coil Capacitance: The pressure coil has capacitancesalong with the inductance. This capacitance increases the power factor of the instrument. Hence causes the error in the reading.
- Error due to Mutual Inductance Effect: The mutual inductance between the pressure and current coil produces an error.
- Eddy Current Error: The eddy current induces in the coil creates its own magnetic field. This field affects the main current flows through the coil. Thus, the error occurs in the reading.
- Stray Magnetic Field: The stray magnetic field disturbs the main magnetic field of the Electrodynamic Wattmeter. Thus, affect their reading.
- Temperature Error: The variation in temperature will change the resistance of the pressure coil. The movement of the spring, which provides the controlling torque also affected because of the temperature change. Thereby, the error occurs in the reading.

The calibration of the electrodynamometer wattmeter is same both for the AC and DC measurement.

References

- Lcr-meter-working-principle-uses: sciencestruck.com, Retrieved 19 March, 2019
- Markoff, john (5 sep 2014). “andrew kay, pioneer in computing, dies at 95”. Obituary. New york times. Retrieved 7 september 2014
- Megger-working-principle-types-history-uses-of-megger: electrical4u.com, retrieved 14 July, 2019
- Electrodynamometer-wattmeter: circuitglobe.com, retrieved 17 May, 2019
- Dunsmore, joel p. (september 2012), handbook of microwave component measurements: with advanced vna techniques, wiley, isbn 978-1-1199-7955-5
- Power-factor-meter: circuitglobe.com, retrieved 19 April, 2019

Permissions

All chapters in this book are published with permission under the Creative Commons Attribution Share Alike License or equivalent. Every chapter published in this book has been scrutinized by our experts. Their significance has been extensively debated. The topics covered herein carry significant information for a comprehensive understanding. They may even be implemented as practical applications or may be referred to as a beginning point for further studies.

We would like to thank the editorial team for lending their expertise to make the book truly unique. They have played a crucial role in the development of this book. Without their invaluable contributions this book wouldn't have been possible. They have made vital efforts to compile up to date information on the varied aspects of this subject to make this book a valuable addition to the collection of many professionals and students.

This book was conceptualized with the vision of imparting up-to-date and integrated information in this field. To ensure the same, a matchless editorial board was set up. Every individual on the board went through rigorous rounds of assessment to prove their worth. After which they invested a large part of their time researching and compiling the most relevant data for our readers.

The editorial board has been involved in producing this book since its inception. They have spent rigorous hours researching and exploring the diverse topics which have resulted in the successful publishing of this book. They have passed on their knowledge of decades through this book. To expedite this challenging task, the publisher supported the team at every step. A small team of assistant editors was also appointed to further simplify the editing procedure and attain best results for the readers.

Apart from the editorial board, the designing team has also invested a significant amount of their time in understanding the subject and creating the most relevant covers. They scrutinized every image to scout for the most suitable representation of the subject and create an appropriate cover for the book.

The publishing team has been an ardent support to the editorial, designing and production team. Their endless efforts to recruit the best for this project, has resulted in the accomplishment of this book. They are a veteran in the field of academics and their pool of knowledge is as vast as their experience in printing. Their expertise and guidance has proved useful at every step. Their uncompromising quality standards have made this book an exceptional effort. Their encouragement from time to time has been an inspiration for everyone.

The publisher and the editorial board hope that this book will prove to be a valuable piece of knowledge for students, practitioners and scholars across the globe.

Index

www.ingramcontent.com/pod-product-compliance
Lightning Source LLC
LaVergne TN
LVHW080702210926
848485LV00003B/406